Engineering
Mega-Systems

The Challenge of Systems
Engineering in the Information Age

COMPLEX AND ENTERPRISE SYSTEMS ENGINEERING

Series Editors: Paul R. Garvey and Brian E. White

The MITRE Corporation

C & E S E www.enterprise-systems-engineering.com

Architecture and Principles of Systems Engineering
Charles Dickerson and Dimitri N. Mavris
ISBN: 978-1-4200-7253-2

Designing Complex Systems: Foundations of Design in the Functional Domain
Erik W. Aslaksen
ISBN: 978-1-4200-8753-6

Engineering Mega-Systems: The Challenge of Systems Engineering in the Information Age
Renee Stevens
ISBN: 978-1-4200-7666-0

Enterprise Systems Engineering: Advances in the Theory and Practice
George Rebovich, Jr. and Brian E. White
ISBN: 978-1-4200-7329-4

Model-Oriented Systems Engineering Science: A Unifying Framework for Traditional and Complex Systems
Duane W. Hybertson
ISBN: 978-1-4200-7251-8

FORTHCOMING

Complex Enterprise Systems Engineering for Operational Excellence
Kenneth C. Hoffman and Kirkor Bozdogan
ISBN: 978-1-4200-8256-2
Publication Date: November 2010

Leadership in Decentralized Organizations
Beverly G. McCarter and Brian E. White
ISBN: 978-1-4200-7417-8
Publication Date: October 2010

Systems Engineering Economics
Ricardo Valerdi
ISBN: 978-1-4398-2577-8
Publication Date: December 2011

RELATED BOOKS

Analytical Methods for Risk Management: A Systems Engineering Perspective
Paul R. Garvey
ISBN: 978-1-58488-637-2

Probability Methods for Cost Uncertainty Analysis: A Systems Engineering Perspective
Paul R. Garvey
ISBN: 978-0-8247-8966-4

Engineering Mega-Systems

The Challenge of Systems
Engineering in the Information Age

Renee Stevens

CRC Press
Taylor & Francis Group
Boca Raton London New York

CRC Press is an imprint of the
Taylor & Francis Group, an **informa** business
AN AUERBACH BOOK

Complex and Enterprise Systems Engineering Series

First Published 2011 by Auerbach Publications

Published 2019 by CRC Press
Taylor & Francis Group
6000 Broken Sound Parkway NW, Suite 300
Boca Raton, FL 33487-2742

© 2011 by Taylor & Francis Group, LLC
CRC Press is an imprint of Taylor & Francis Group, an Informa business

No claim to original U.S. Government works

ISBN 13: 978-1-4200-7666-0 (hbk)

Library of Congress Cataloging-in-Publication Data

Stevens, Renee.
   Engineering mega-systems : the challenge of systems engineering in the information age / Renee Stevens.
      p. cm. -- (Complex and enterprise systems engineering)
   Includes bibliographical references and index.
   ISBN 978-1-4200-7666-0 (hardcover : alk. paper)
   1. Systems engineering. 2. Large scale systems. I. Title. II. Series.

   TA168.S72 2010
   620.001'171--dc22   2010012611

Visit the Taylor & Francis Web site at
http://www.taylorandfrancis.com

and the CRC Press Web site at
http://www.crcpress.com

Engineering Mega-Systems

The Challenge of Systems Engineering in the Information Age

Renee Stevens

CRC Press
Taylor & Francis Group
Boca Raton London New York

CRC Press is an imprint of the
Taylor & Francis Group, an **informa** business
AN AUERBACH BOOK

Complex and Enterprise Systems Engineering Series

First Published 2011 by Auerbach Publications

Published 2019 by CRC Press
Taylor & Francis Group
6000 Broken Sound Parkway NW, Suite 300
Boca Raton, FL 33487-2742

ISBN 13: 978-1-4200-7666-0 (hbk)

Library of Congress Cataloging-in-Publication Data

Stevens, Renee.
    Engineering mega-systems : the challenge of systems engineering in the information age / Renee Stevens.
        p. cm. -- (Complex and enterprise systems engineering)
    Includes bibliographical references and index.
    ISBN 978-1-4200-7666-0 (hardcover : alk. paper)
    1. Systems engineering. 2. Large scale systems. I. Title. II. Series.

TA168.S72 2010
620.001'171--dc22
                                                    2010012611

Visit the Taylor & Francis Web site at
http://www.taylorandfrancis.com

and the CRC Press Web site at
http://www.crcpress.com

Dedication

To Bill for his encouragement, his patience and, most of all, his love.

Contents

SECTION II CONCEPTS AND FRAMEWORKS

SECTION III   CASE STUDIES IN ENGINEERING MEGA-SYSTEMS

List of Figures

List of Tables

List of Acronyms

A

ACMC: Assistant Commandant, U.S. Marine Corps
ADSI: Air Defense System Integrator
ADSIM: Air Defense Simulation
AEGIS: Aegis combat system is an integrated weapons system used by the U.S. Navy
AFI: Air Force Instruction
AFIT: Air Force Institute of Technology
AFPD: Air Force Policy Directive
ALE: Application Level Events
APM: Association for Project Management
ARC: Architecture Review Committee
ARPA: Advanced Research Projects Agency
ARPANET: Advanced Research Projects Agency Network
ARV Aslt: Armed Robotic Vehicle Assault
ARV RSTA: Armed Robotic Vehicle Reconnaissance Surveillance and Target Acquisition
ASCIET: All-Service Combat Identification Evaluation
ASL: Action Semantic Language
AT&T: American Telephone and Telegraph Company
ATM: Asynchronous Transfer Mode
Auto-ID: Automatic IDentification
AWACS: Airborne Warning and Control System

B

BMC2: Battle Management Command and Control
BMD: Ballistic Missile Defense
BMDO: Ballistic Missile Defense Organization
BMDS: Ballistic Missile Defense System
BUG-E: Battlefield Universal Gateway Equipment

C

C2V: Command and Control Vehicle
C4ISR: Command, Control, Communications, Computers, Intelligence, Surveillance and Reconnaissance
CA/T: Central Artery/Tunnel
CAOC: Combined Air Operations Center
CapWIN: Capital Wireless Integrated Network
CASPIAN: Consumers Against Supermarket Privacy Invasion and Numbering
CDC: Centers for Disease Control and Prevention
CDD: Capability Development Document
CE: Chief Engineer
CERN: European Organization for Nuclear Research
CIO: Chief Information Officer
CIS: Combat Intelligence System
CJCS: Chairman, Joint Chiefs of Staff
CJCSM: Chairman, Joint Chiefs of Staff Manual
CNN: Cable News Network
COT: Cursor On Target
COTS: Commercial-Off-The-Shelf
CRM: Customer Relations Management
CSTB: Computer Science and Telecommunications Board
CTAPS: Contingency Theater Advanced Planning System
CV: Carrier Vehicle

D

DAU: Defense Acquisition University
DCGS: Distributed Common Ground System
DDS: Distributed Denial of Service
DEP: Distributed Engineering Plant
DHS: Department of Homeland Security
DIB: DCGS Integration Backbone
DISA: Defense Information Systems Agency
DNS: Domain Name Service
DoD: Department of Defense
DoDAF: Department of Defense Architecture Framework

E

EA: Enterprise Architecture
EAN: European Article Numbering
e-Government: Electronic Government
EPC: Electronic Product Code

EPCIS: Electronic Product Code Information Services
EPLRS: Enhanced Position Location Radio Systems
ERP: Enterprise Resource Planning
EWG: European Working Group

F

FAA: Federal Aviation Administration
FAST: Free and Secure Trade
FCS: Future Combat Systems
FDA: Food and Drug Administration
FEA: Federal Enterprise Architecture
FNC: Federal Networking Council
FOC: Full Operational Capability
FRMV: FCS Recovery and Maintenance Vehicle
FY: Fiscal Year

G

GCCS: Global Command and Control System
GM: General Motors
GPRS: General Packet Radio Service
GPS: Global Positioning System

H

HF: High Frequency
HTML: Hypertext Markup Language
HTTP: Hypertext Transport Protocol
HWIL: Hardware-In-the-Loop

I

IABM: Integrated Architecture Behavior Model
IAG: Industry Action Group
IBuild: JDEP Infrastructure Build
ICP: Interface Change Proposal
IEEE: Institute of Electrical and Electronics Engineers
IETF: Internet Engineering Task Force
INCOSE: International Council on Systems Engineering
IP: Internet Protocol
IRS: Internal Revenue Service
ISAC: Information Sharing and Analysis Center
ISE: Information Sharing Environment
ISO: International Standards Organization

ISO/IEC: International Standards Organization/International Electrotechnical Commission
ISR: Intelligence, Surveillance, and Reconnaissance
IT: Information Technology

J

JCIDS: Joint Capabilities Integration and Development System
JCIET: Joint Combat Identification Evaluation Team
JDEP: Joint Distributed Engineering Plant
JDN: Joint Data Network
JEZ: Joint Engagement Zone
JFCOM: Joint Forces Command
JINTACCS: Joint Interoperability of Tactical Command and Control Systems
JMETC: Joint Mission Environment Test Capability
Joint STARS: Joint Surveillance Target Attack Radar System
JRG: Joint Requirements Group
JROC: Joint Requirements Oversight Council
JSPC: Joint Strategy and Planning Committee
JSSEO: Joint SIAP System Engineering Organization
JTAMD: Joint Theater Air and Missile Defense
JTAMDO: Joint Theater Air and Missile Defense Organization
JTIDS: Joint Tactical Information Distribution System

L

LAN: Local Area Network
LinX: Law Enforcement Information Exchange

M

MA-SHARE: Massachusetts Simplifying Healthcare Among Regional Entities
MCS: Mounted Combat System
MDA: Missile Defense Agency
MDA®: Model-Driven Architecture
MDWAR: Missile Defense Wargame Analysis Resource
MIDS: Multifunction Information Distribution System
MIT: Massachusetts Institute of Technology
MV-E: Medical Vehicle Evacuation
MV-T: Medical Vehicle Treatment

N

NASA: National Aeronautics and Space Administration
NASCIO: National Association of State Chief Information Officers

NATO: North Atlantic Treaty Organization
NCIS: Naval Criminal Investigative Service
NCSA: National Center for Supercomputing Applications
NCW: Network Centric Warfare
NECC: Net-Enabled Command Capability
NIMDA: "admin" spelled backward; the name of a computer worm
NLOS-C: Non-Line-of-Sight Cannon
NLOS-LS: Non-Line-of-Sight Launch System
NLOS-M: Non-Line-of-Sight Mortar

O

ODDSCAPE: Operational Data Driven Simulation for Correlation Algorithm Performance Evaluation
OMB: Office of Management and Budget
OMG: Open Management Group
ONS: Object Name Service
OSAF: Office of the Secretary of the Air Force
OSD: Office of the Secretary of Defense

P

PIM: Platform Independent Model
PLC: Public Limited Company
PLGR: Precision Lightweight GPS Receiver (See GPS)
PM: Program/Project Manager
PMBOK: Project Management Book of Knowledge
PMI: Project Management Institute
PML: Physical Markup Language
POET: Political, Operational, Economic and Technical
POS: Point Of Sale
PSM: Platform Specific Model

R

RAF: Royal Air Force
RDD: Requirements Description Document
RF: Radio Frequency
RFID: Radio Frequency Identification
RMA: Revolution in Military Affairs
RSV: Reconnaissance and Surveillance Vehicle

S

SADL: Situational Awareness Data Link
SAGE: Semi-Automatic Ground Environment

SAP AG: SAP AG is a developer of enterprise software
SBI: Department of Homeland Security's Secure Border Initiative
SBInet: The technology portion of SBI
SCO: Screening Coordination Office
SDC: Strategic Direction Committee
SE: Systems Engineering
SE: System Engineering (as used by SIAP)
SENTRI: NEXUS/Secure Electronic Network for Travelers Rapid Inspection
SETF: Systems Engineering Task Force
SIAP: Single Integrated Air Picture
SIAP SETF: SIAP System Engineering Task Force
SIPRnet: Secret Internet Protocol Router Network
SOF: Special Operations Forces
SOSCOE: System of Systems Common Operating Environment

T

TACOM: Tank-Automotive and Armaments Command
TAG: Technical Action Group
TAMD: Theater Air and Missile Defense
TBMCS: Theater Battle Management Core Systems
TCP: Transmission Control Protocol
TCP/IP: Transmission Control Protocol/Internet Protocol
TCTF: Time Critical Targeting Functionality
TTPs: Tactics, Techniques, and Procedures
TWIC: Transportation Worker Identification Credential

U

UCC: Uniform Code Council
UCORE: Unified Core
UGV: Unmanned Ground Vehicle
UHF: Ultrahigh Frequency
U.K.: United Kingdom
U.S.: United States
USS: United States Ship
UML: Unified Modeling Language
URL: Uniform Resource Locator
US-VISIT: United States Visitor and Immigrant Status Indicator Technology
USAF: United States Air Force
USCENTCOM: United States Central Command
USN: United States Navy

V

VADM: Vice Admiral

W

W: Watt
W3C: World Wide Web Consortium
WAP: Wireless Application Protocol
WCCS: Wing Command and Control System
WiFi: Wireless Fidelity
WWW: World Wide Web

X

XML: eXtensible Markup Language

Acknowledgments

This book was researched and written under the auspices of The MITRE Corporation—a private, independent, not-for-profit organization chartered to work in the public interest. MITRE technical staff are experts who apply systems engineering and research and development expertise to critical issues of national importance in a variety of advanced technological and scientific fields. MITRE currently manages federally funded research and development centers for the DoD, the Federal Aviation Administration, the Internal Revenue Service, and the Department of Homeland Defense.

To define where the company will focus its energy, resources, and skills in the coming years, MITRE establishes a set of corporate goals. Each of these goals drives one or more initiatives identified by teams of officers to bring MITRE closer to its goal. One of these initiatives explores the boundaries within which traditional systems engineering is effective and undertakes a survey of approaches that may be successful outside these boundaries. This book represents one of many activities in support of the goal of developing, describing, and validating engineering methods that address systems beyond the reach of traditional systems engineering.

Many individuals have contributed to the realization of this book, and to the evolution and refinement of the ideas presented here. The original impetus came from a challenge by Alfred Grasso, then Senior Vice President of MITRE, to synthesize lessons learned about the engineering of systems and to turn those lessons into a published book. What had started out (perhaps naively on my part) as a straightforward effort has, over the course of months of considerable research and—even more importantly—enriching dialogue with my colleagues, evolved into a view of an emerging set of engineering opportunities and challenges. Throughout the process, Grasso has urged me on, providing strong support and encouragement. Others at MITRE have also furthered this effort: Ray Haller, Jason Providakes, and Peter Sherlock have provided me with invaluable advice and encouragement, as well as funding that enabled me to devote some portion of my regular workweek to this effort. Together, they have given me the opportunity and the means that have resulted in this book.

Many colleagues allowed me to share my evolving ideas with them, and their thoughtful and thought-provoking comments were key to refining those ideas. Dave Alberts, Chuck Howell, Mike Kuras, Mike Lavine, George Providakes, Peter Sharfman, and Brian White are among those who took the time to help me frame my ideas and who offered early and invaluable reviews.

I am particularly grateful to the many people who provided information on which to build the case studies and who shared their recollections, files, and insights with me. Richard Staats drafted the illustrative example of the value of mega-systems found in Chapter 1. Tom Nyman shared his extensive notes and recollections about the early phases of the Single Integrated Air Picture (SIAP) project, and Karen Rigopoulos and Kim Crider contributed their more recent experiences and insights. Captain Jeffery Wilson, Technical Director of the Joint SIAP Systems Engineering Organization, reviewed the final draft and encouraged its publication. David Brock of MIT provided me with invaluable insight into the origins of the Auto-ID technologies.

To those at MITRE and outside who served as peer reviewers, I owe special thanks. In particular, I wish to mention Donna Rhodes and Tom Hughes. Their experience, insight, and perspective contributed immeasurably to this final product. Their confidence that this material was worthy of publication was particularly encouraging.

I would like to acknowledge with special thanks the services of Margaret MacDonald, who helped me edit this book. She was critical in helping me convey my intended meaning and make it seem easy to do.

Finally, while I relied on many sources both inside MITRE and in the public domain, the observations and conclusions are mine, and I take full responsibility for them.

About the Author

Renee G. Stevens is a Senior Principal Engineer at The MITRE Corporation. She has had 30 years of experience in the analysis, engineering, and acquisition of large-scale systems, primarily for the Department of Defense and other government agencies. Her current interests lie in research and practice contributing to the development of an enterprise systems engineering discipline.

Stevens has developed a well-received Profiler for use in characterizing the environment and context in which a system will be developed and will operate. It serves as both a diagnostic tool and the basis for a situational model. Results have been widely briefed to government, academic, and professional audiences. She is applying the model to the assessment of several large-scale programs and is conducting research on innovative strategies and practices to improve the acquisition of information technology systems.

Stevens received her bachelor's degree in political science from Hunter College, City University of New York, in 1966, and a master's degree in public and business administration from George Washington University in 1981. She is a member of the Institute of Electrical and Electronic Engineers and the Academy of Management.

SETTING
THE STAGE

1

Chapter 1

Introduction

1.1 The Trend Toward Large-Scale, Richly Interconnected Systems

Government agencies, both military and civilian, and indeed the global business community as a whole, are moving aggressively to leverage and capitalize on the advances of information technologies. Not only have these technologies provided revolutionary new capabilities, but also they have stimulated fundamental changes in how organizations, including military units, accomplish their tasks and achieve their objectives. This, then, is transformation.[1] Transformation is not about doing the same things better, but about leveraging new technologies to accomplish the same missions in fundamentally different ways.

> Mega-systems are defined as "those large-scale, complex systems that cross traditional boundaries to provide a level of functionality not achieved by their component elements."

Transformation is occurring not only in the military, but also in government and private sectors. Governments are relooking at historical ways of doing business and leveraging information technologies to do so. They are establishing shared service agreements with other agencies and are forming public–private partnerships. They are transforming how services are delivered to citizens, businesses, and taxpayers. The same transformation patterns are evident in the business sector. There, technology is being used to improve performance, enhance customer relations, and collaborate with strategic partners across the global supply chain. In all these instances, information that once was resident within a single function, agency, or corporation is being shared with others.

The move toward cross-boundary solutions, enabled by Internet technologies, is broad based. This book uses defense mega-systems as a primary example, not because this book is specifically about the Department of Defense (DoD), but because the department has had the most experience in developing such systems. Other government agencies, such as the Department of Homeland Security, the Federal Bureau of Investigation, or the Federal Aviation Administration, as well as the business community in general can learn from the experiences of the DoD and, similarly, the DoD can gain insight from the experience of other agencies and the commercial sector.

Within the DoD, transformation arises from a number of converging trends. Changes in the strategic environment and a broader, more uncertain threat context make transformation necessary; the emergence of highly capable information technologies makes it possible. In response to these external factors, the DoD is altering its institutional environment (how it intends to do business) as well as how it intends to fight. The expected outcome is an agile, richly interconnected force that can orchestrate available capabilities in previously unanticipated ways. This underscores a trend away from a reliance on stand-alone component systems—often referred to as a platform-centric approach—to the creation of increasingly interdependent systems that cross traditional boundaries. These large-scale, complex systems cross traditional boundaries to provide a level of functionality not achieved by their component elements. We call them *mega-systems*.

In other parts of the federal government, agencies confront the daunting challenge of integrating systems that have been developed and operated separately, yet need to work effectively together to accomplish overarching missions. This is perhaps most striking in the Department of Homeland Security (DHS), which is working to transform 22 legacy agencies into one department and to collaborate with other federal agencies and with state and local governments, as well as with private interests.

In the commercial world, companies form mega-systems not only as they establish enterprisewide systems to link employees and processes internally, but also as they extend them to connect partners, customers, and suppliers—in effect, "outsiders." The emerging concept is that of the "extended enterprise." In this concept, participants are linked on the basis of their role rather than their geographic location or business affiliation, and "just-in-time" processes supplant traditional batch processes or near-real-time information systems. Here, no single company owns the entire suite of hardware, software, and data. Rather, they are intermingled into an intricate, interconnected, and secure network.

We build mega-systems because we expect that they will yield substantial operational benefits. As in the DoD, where joint operations are expected to yield a competitive military advantage, the expected outcome in the commercial world is a competitive business advantage.

THE BENEFITS OF MEGA-SYSTEMS:
AN ILLUSTRATIVE EXAMPLE

Connecting several independent systems into a mega-system offers several advantages. In the case of this illustrative example, the integration of separate component systems into an air defense mega-system offers substantial increases in *persistence, range, synergy,* and *agility.*

Mega-systems tend to have *increased persistence* over their component systems. Persistence has two characteristics: survival and staying power. Because a mega-system is composed of independent systems, these component systems can have significantly different survival characteristics in varying environments. While some component systems may fail, others can survive and guarantee mission accomplishment. It is not just the physical characteristics of the component systems that offer this increased survivability: geographical and temporal dispersion of the mega-system's components also increase survivability. Synergistic interactions among the component systems in a mega-system can increase the length of time over which a mega-system can be active and effective.

In the air defense example, the active radar systems can take turns actively tracking, thereby decreasing the total chance that the enemy will detect the individual radars and increasing the survivability of the mega-system. Likewise, the combination of active and passive sensors can potentially allow some sensors to survive even if others are vulnerable to blinding or detection and destruction. Thus, even if the air defense mega-system loses a significant number of individual sensor systems, the mega-system may remain fundamentally mission effective.

The mega-system also offers an *increase in range or spectrum* over its component systems. By range we mean the number of potentially simultaneous actions that can be taken in the attempt to accomplish a mission. The mega-system is designed or is "grown" with a particular purpose or mission in mind. By definition, the component systems in a mega-system are capable of acting independently. This can increase the probability of mission success by pursuing the mission using independent means that are effective in different operating environments.

In the air defense example, the system may include a combination of shoulder-fired missiles, mounted medium- and high-altitude missiles, kinetic energy weapons (e.g., guns), and directed energy weapons (e.g., lasers). No individual tactic or technological countermeasure could easily defeat all of these means of attack, whereas an enemy airframe would have a better chance of defeating the component systems individually.

(continued on next page)

Mega-systems combine the capabilities of component systems into *synergistic capabilities.* One of the most prominent examples of this is increased reach (when and where the mega-system can act) compared to the individual systems. In the case of the air defense mega-system, the high-, medium-, and low-altitude missiles provide comprehensive, overlapping protection against enemy airframes flying from ground level to 60,000 feet or more above the earth.

Evidence emerging from both the military and commercial worlds indicates that these benefits are, in fact, being realized. In the illustrative air defense example above, the DoD expects to gain these advantages in terms of enhanced persistence, range, reach, and agility. An analogous commercial example might describe the benefits that would accrue from information sharing across the supply chain in terms of reduced cycle times, reduced inventory levels, improved sales, and improved customer services.

1.2 Why This Book?

This book explores the engineering and acquisition of this evolving class of systems. It argues that the traditional approaches to large-scale systems engineering, and the accompanying acquisition processes that have developed in the past, are inadequate for the development and evolution of effective mega-systems.

For several reasons, including their sheer scale, the nature and pace of change of their underlying technologies, the potential complexity of their interactions, and—perhaps most importantly—the fact that a single organization rarely owns and therefore completely controls the mega-system, engineering these mega-systems entails new challenges. The fundamental question that we must ask ourselves is: To what extent and under what circumstances do the practices and processes of systems engineering that evolved in the post-World War II era continue to apply to these massively interconnected, information technology-intensive mega-systems? A clearly related question is: Given that there are circumstances in which traditional approaches may no longer apply, what new practices and processes might be required?

The core of the problem is that traditional systems engineering rests upon the careful specification of a detailed and internally consistent set of requirements. The requirements for the system as a whole flow down into requirements for the various subsystems, requirements for characteristics of the materials used in constructing hardware, interface and runtime requirements for software, and requirements for the supporting infrastructures for supply, maintenance, and training. The ideal state for the systems engineer is a set of requirements that remain stable throughout the design and construction of the system. Because this is rarely possible in the real world, systems engineers use a highly disciplined process to identify, track, and implement all the consequences of the unavoidable minimum of changes in requirements.

In addition, government agencies, such as the DoD and the National Aeronautics and Space Administration (NASA), have built their acquisition process around the specification of requirements. For example, in the DoD, users interact with the procurement system in that they (or their representatives) establish requirements that determine what hardware and software will be designed, built, and procured. The basis of competition to sell goods and services is a set of requirements sufficiently precise and comprehensive to allow objective comparison of the alternative vendors.

Unfortunately, it is difficult, if not impossible, to establish a clear and precise set of requirements for a typical mega-system. This is not because mega-systems are large and complicated systems of systems. For example, in the past the DoD has been able to define the requirements for something as complex as a ballistic missile nuclear submarine, encompassing a wide range of cutting-edge technologies that had to work together with extremely narrow margins of error. Once the requirements had been established, the systems engineering (e.g., designing a nuclear warhead that would fit within a ballistic missile that would fit inside the submarine) was straightforward, although clearly technically challenging.

Mega-systems are different from traditional systems because:

- ▪ The systems of which they are composed are purposeful in their own right. That is, each fulfills a role independent of its role in the larger mega-system.
- ▪ A typical component system may be a critical part of several, quite different mega-systems, each of which imposes a different set of priorities.
- ▪ Component systems are often under the control of different organizations, each with its own set of priorities and constraints.
- ▪ The component systems will likely be designed and acquired on different timetables.

At the same time, the systems engineers for each component must accept that it is impossible to meet the requirements of all the mega-systems in which their individual systems will participate. Moreover, systems engineers have neither the insight nor the authority to make trade-offs among the requirements imposed by different mega-systems.

In short, the design and acquisition of a mega-system must, by its very nature, violate the two basic principles of traditional systems engineering and acquisition: (1) clarity of requirements and discipline in changing them; and (2) an established hierarchy, with someone clearly in charge who has both the duty and the authority to make trade-offs and decisions.

How, then, can any enterprise meet the challenge of developing and evolving the systems it deems critical to the transformation of its organization and the achievement of its mission?

This book approaches that question from both the conceptual and the practical perspectives. The conceptual perspective provides a vocabulary and a framework with which to explore issues relevant to the characteristics that differentiate mega-systems from traditional, well-bounded systems. That framework then evolves into a Profiler that can be used both to understand the nature and context of the system at hand and, on that basis, to select the most appropriate processes, tools, and techniques. The practical perspective allows us to leverage experiences in the engineering and acquisition of systems that approximate the understanding of mega-system characteristics and issues. We look to the experience of both the commercial world and the world of civil governance for insights on how to define problems connected to mega-systems and where to seek solutions.

In this context, we examine two case studies to identify lessons learned and *emerging first practices*. It is not reasonable or even possible to identify best practices at this early state; that will take considerably more time and a broader set of case studies.

The first case study traces the development of the Single Integrated Air Picture (SIAP) within the DoD (see Chapter 7). This effort was initiated to solve long-standing problems related to track data for air and missile defense applications. While the participating systems all used the same data link standards, their individual implementations were sufficiently different that the separate results could not be readily integrated. The SIAP program is not responsible for developing the systems themselves, but for developing common solutions that each system would integrate separately. While this is by no means a typical DoD acquisition effort, it does span multiple formal acquisition programs and highlights the challenges of developing and implementing solutions that cross program boundaries.

The second case describes the efforts to develop a common Radio Frequency Identification (RFID) capability for use across the global supply chain to track products from manufacture, through distribution, to sale, and potentially to final disposal (see Chapter 8). This effort was sponsored by a consortium of commercial suppliers, retailers, and vendors, and highlights the development of an extended enterprise capability.

These two efforts are exploring new organizational constructs and new engineering approaches. They permit a retrospective look at lessons learned to date and a prospective look at lessons still to be learned.

An important caution is warranted here. The study of mega-systems reported in this book is by no means intended to be definitive. Rather, it builds on the conceptual groundwork in system-of-systems engineering, complex adaptive systems, and network-centric operations, as well as on practical experience in actually engineering and acquiring these large-scale, complex systems. In doing so, the hope is to identify some key tenets and first practices related to both the design and development of such systems. Even more important, the aim is to initiate a rich dialogue not only within the systems engineering community—both researchers and practitioners—but also among those government leaders responsible for directing, executing, and overseeing the development of mega-systems.

1.3 Organization of the Book

This book proceeds from broad, conceptual foundations to more specific topics and concludes with some recommendations for charting the way ahead. It is organized in four sections.

Section I sets the stage. Chapter 2 provides the national security context and briefly explores the trends leading government agencies (including the DoD) and the private sector to embark upon the development of the large-scale, massively interconnected systems that we call mega-systems. Of particular interest are changes in the strategic and operational context from a period of relative predictability to an era of uncertainty regarding the source and nature of threats. These changes, in conjunction with trends in information technologies, underpin the "emerging American way of war," also termed *network-centric warfare*. Related to these trends are fundamental changes in the institutional processes by which the DoD defines needed capabilities and acquires systems.

Section II of this book provides the conceptual underpinnings, with references to insights that can be drawn from the civilian world, both governmental and private. Chapter 3 proposes a definition of mega-systems and discusses the different ways in which they may emerge. Chapter 4 then offers a framework for understanding these systems that takes into account their technical characteristics and behavior, the environment in which they are expected to operate, and the stakeholder context. Chapter 5 identifies some of the known challenges in engineering and acquiring mega-systems, and discusses the relevant systems engineering processes. It introduces a Profiler for use as a self-assessment tool to help in characterizing the nature and context of the system of interest and also as the basis of a situation model that could help systems engineers select and adapt processes, tools, and techniques to the circumstances at hand.

Section III explores the practical aspects of engineering and acquiring mega-systems through two case studies. Key to Section III are summaries of the emerging lessons learned and emerging first practices associated with these particular efforts.

Section IV then seeks to synthesize the conceptual and the practical aspects by developing a set of applicable tenets and proposing a way ahead.

Endnote

1. Within the DoD, "transformation" is defined as "a process that shapes the changing nature of military competition and cooperation through new combinations of concepts, capabilities, people and organizations that exploit our nation's advantages and protect against our asymmetric vulnerabilities to sustain our strategic position, which helps underpin peace and stability in the world." *Transformation Planning Guidance*, April 2003.

Chapter 2

Context and Trends

The United States is transitioning from an industrial age to an information age military. This transition requires transformation in warfighting and the way we organize to support the warfighter. Although the end-state of transformation cannot be fully defined in advance, we do know some of the necessary prerequisites for transformation. In particular, we know that early transformation requires exploiting information technology to reform defense business practices and to create new combinations of capabilities, operating concepts, organizational relationships and training regimes.

—Transformation Planning Guidance, April 2003

Several factors combine to fundamentally change the nature of the systems developed and fielded not only to military forces but also to civilian government agencies and to the commercial world. Like many other military forces, the U.S. military faces a strategic environment that requires agile and adaptive response to a wide range of threats and missions. Responding to this uncertainty is the emerging concept of network-centric warfare, which seeks to leverage information as a competitive source of power. Other government agencies and private industry also need the ability to respond with agility and flexibility to unexpected demands and opportunities. The information revolution provides the tools to interconnect a wide range of elements and provide them timely information. Finally, there are significant changes in the processes by which all organizations intend to acquire capabilities. Together, these conditions lead to growing opportunities for large-scale, richly interconnected systems that bridge traditional organizational and functional boundaries.

This chapter briefly outlines the converging trends influencing formal system acquisition by government agencies as well as system development in the private sector. At its conclusion, the "mega-system" is presented as the next-generation challenge for systems engineering and contrasts the emerging engineering process associated with mega-systems with the processes developed in the latter half of the twentieth century.

2.1 Changing the Strategic Environment in the U.S. Department of Defense

The DoD has clearly recognized the need to transition from the Cold War environment, in which the United States faced a primary, well-studied adversary, to today's situation, which presents a wide range of potential hostile environments, military operations, and possible scenarios. Rooted in fundamental political, socio-cultural, and economic forces, the types of threats we face have become substantially broader. While we may still need to confront traditional nation-states, we must also be prepared to deal with diverse transnational "ad hoc" terrorist groups. The nature of the threat has also broadened. Conventional military operations remain a menace; however, asymmetric threats are growing as our enemies seek to exploit weaknesses of American military forces or to indiscriminately target noncombatants. The United States has become a prime target, and operations abroad can no longer be isolated from potential attacks on the homeland or against American interests abroad.

Key to the ability to respond to this environment of uncertainty and complexity is the need for agility and adaptability. The long cycle times, well-developed tools and processes, and deliberate planning that characterized the industrial age must be replaced with short cycle times, new capabilities, and adaptive planning. Similarly, operations involving multiple branches of the military Services were, in the past, designed around separate missions, sequential phases, and well-defined geographical responsibilities, and, unfortunately, were often characterized by "tortured interoperability." Here, interoperability is formally defined as the "ability of systems, units or forces to provide services to and accept services from other systems, units or forces and to use the services so exchanged to enable them to operate effectively together." The phrase "tortured interoperability," taken from briefings and writings of the late Vice Admiral Arthur K. Cebrowski, at the time Director of the DoD Office of Force Transformation, refers to the difficulties that have been encountered in sharing information among the separate Services. In contrast, the desired goal state entails more fully integrated and, in some case, interdependent joint operations.

Many of these characteristics of the emerging American way of war were evident in the early days of Operations Enduring Freedom and Iraqi Freedom. In prepared testimony[1] before the Senate Armed Services committee on July 9, 2003, Secretary of Defense Donald Rumsfeld and General Tommy Franks, Commander, Central Command, presented the "key lessons so far":

- The importance of *speed*, and the ability to get inside the enemy's decision cycle and strike before he is able to mount a coherent defense
- The importance of *jointness*, and the ability of U.S. forces to fight, not as individual de-conflicted Services, but as a truly joint force—maximizing the power and lethality they bring to bear
- The importance of *intelligence*, and the ability to act on intelligence rapidly—in minutes instead of days or even hours
- The importance of *precision*, and the ability to deliver devastating damage to enemy positions while sparing civilian lives and the civilian infrastructure

These operational themes are increasingly tightly intertwined.

Cordesman (2003), in his *Instant Lessons of the Iraq War*, pointed out the relationship between joint operations and tempo:

> The U.S. had an almost incredible advantage in terms of its ability to bring together land and air operations and support them from the sea and friendly bases at very high tempos of coordinated operations and shift the mix of joint operations according to need over the entire theater of operations. The issue was far more than Jointness per se; it was the coordination and sheer speed of operations at every dimension of combat.

Flexibility—the ability to rapidly compose forces that lack a habitual command relationship and to combine capabilities in ways tailored to the circumstances at hand—and agility—the ability to rapidly respond to opportunities and transition between tasks—were also hallmarks of these operations. General Tommy Franks, commander of U.S. Central Command (USCENTCOM), described these complex joint interdependencies in a March 22, 2003, briefing[2] on military operations from his USCENTCOM headquarters, as follows:

> Let me talk for a minute about our capabilities. The coalition now engaged in and supporting Operation Iraqi Freedom includes Army and Marine forces from the land component; air forces from several nations; naval forces, to include the Coast Guard; and Special Operations forces.
>
> Our plan introduces these forces across the breadth and depth of Iraq, in some cases simultaneously and in some cases sequentially. In some cases, our Special Operations forces support conventional ground forces. Examples of this include operations behind enemy lines to

attack enemy positions and formations, or perhaps to secure bridges and crossing sites over rivers, or perhaps to secure key installations, like the gas-oil platforms, and, of course, in some cases, to adjust air power, as we saw in Afghanistan.

Now, in some cases, our air forces support ground elements or support special operations forces by providing (inaudible) and intelligence information, perhaps offensive electronic warfare capabilities. At other times, coalition airmen deliver decisive precision shock, such as you witnessed beginning last night.

At certain points, special operations forces and ground units support air forces by pushing enemy formations into positions to be destroyed by air power. And in yet other cases, our naval elements support air, support ground operations or support Special Operations forces by providing aircraft, cruise missiles or by conducting maritime operations or mine-clearing operations.

And so the plan we see uses combinations of these capabilities that I've just described. It uses them at times and in places of our choosing in order to accomplish the objectives I mentioned just a moment ago.

That plan gives commanders at all levels and it gives me latitude to build the mosaic I just described in a way that provides flexibility so that we can attack the enemy on our terms, and we are doing so.

These emerging characteristics—richly networked joint and coalition forces, capable of operating at high tempos and able to adapt to and leverage opportunities as they emerge—are hallmarks of the emerging future force.

The emerging concept of network-centric warfare/operations, defined as military operations that are enabled by the networking of the force,[3] provides the context to understand the expected increases in combat power resulting from a richly connected, information-enabled force. The term was first widely introduced in an article entitled "Network Centric Warfare: Its Origins and Future" (Cebrowski and Garstka, 1998), which proposed a way of thinking about warfare in the information age that parallels and leverages the very technologies and processes that have changed American business. The themes that provide a competitive business advantage to the American and global economy are also expected to provide a competitive military advantage when applied to the conduct of warfare. These themes are:

- The shift in focus from the platform to the network
- The shift from viewing actors as independent to viewing them as part of a continuously adapting ecosystem
- The importance of making strategic choices to adapt or even survive in such changing ecosystems

At the most fundamental level, network-centric concepts, whether in the commercial or military domain, share the fundamental hypothesis that not merely information, but *shared* information, represents a source of potential value and competitive advantage. In business, such value is measured in terms of functionality, reliability, convenience, and cost. In the military domain, particularly in combat operations, value is measured using terms such as survivability, lethality, speed, timeliness, and responsiveness.

Network-centric warfare embodies a set of concepts that allows warfighters to take full advantage of all available information and to bring all available assets to bear in a rapid and flexible manner. It is based on four fundamental, linked tenets (Alberts and Hayes, 2002):

1. A robustly networked force improves information sharing and collaboration.
2. Information sharing and collaboration enhance the quality of information and shared awareness of the situation.
3. Shared situation awareness enables self-synchronization.
4. Shared synchronization dramatically increases mission effectiveness.

The theories underlying network-centric warfare give rise to a view of battle management and command and control that supports this highly networked, agile force and provides all appropriate entities with the information necessary to sense, understand, decide, and act. For the forces to be agile, the battle management and command and control capability that supports them must also be agile. It can no longer be specific to a particular force element, its well-established command relationships, and its particular concepts of operations. Rather, such an agile capability must accommodate and adapt to the changing and often uncertain operating environment. It must allow the rapid and effective sharing of information across what is emerging as the "extended military enterprise": the situation-specific mix of U.S. and coalition forces and non-government organizations. This translates into capabilities to enable:

■ Superior decision making across multiple dispersed and distributed command elements, including those of multinational forces
■ Shared understanding among all friendly elements on the battlefield, including those of non-governmental organizations, which can include such diverse organizations as the Red Cross and Doctors Without Borders
■ The ability to support tailorable organizations that may be constructed dynamically to accomplish a particular mission
■ Operation-wide integration, including access to all weapons authorized to accomplish the mission

Underlying these capabilities is an infrastructure with attributes that include universal communications—defined as available and trustworthy means to communicate with any authorized person, system or weapon anywhere—secure information, data interoperability, and command and control systems that are easy to use. That infrastructure, when coupled with the necessary culture, people, and training, is what yields the desired outcome: significantly enhanced mission effectiveness.

2.2 The Imperative to Share Information across Agencies

Just as the military departments within the DoD recognize that sharing information increases their overall effectiveness, federal agencies, as well as state and local governments, have recognized that information sharing is vital not only to continuing anti-terrorism efforts, but also to delivering services to citizens, preparing for and responding to natural disasters, and providing for public health and public safety. In all these cases, sharing information means crossing organizational boundaries to gain access to information that may initially have been collected for different purposes and invariably has been stored in different repositories, using different formats. In some cases, it means bridging these separate stovepipe systems; in others, it means building new systems that, from the onset, provide the means to share information between agencies and across different levels of government.

2.2.1 Information Sharing to Counter Terrorism

In the Executive Summary of *The 9/11 Commission Report*, the members of the Commission note that one of the key elements in achieving unity of effort across the U.S. Government is "unifying the many participants in the counterterrorism effort and their knowledge in a network-based information sharing system that transcends traditional government boundaries" (National Commission on Terrorist Attacks Upon the United States, 2004). The Commission went on to recommend a government-wide effort to solve the legal, policy, and technical issues necessary to create a "trusted information network" that not only bridges the member agencies of the intelligence community, but also extends the network to other public agency and relevant private-sector databases.

In December 2004, the U.S. Congress passed the Intelligence Reform Act (Intelligence Reform and Terrorism Prevention Act of 2004), which calls for the establishment of an Information Sharing Environment (ISE) to link people, systems, procedures, and technologies across federal, state, local, and tribal entities and with the private sector. The National Strategy for Information Sharing was released in October 2007 and called for the establishment of a network of state and major urban area fusion centers (White House, 2007). The strategy envisioned these fusion centers as the focus, although not the exclusive points, for sharing terrorism information, homeland security information, and terrorism-related law enforcement information with state and local governments. Building such a network is not only a significant technical undertaking, but also requires individual organizations to reexamine their traditional patterns of interaction, trust, competition, and cooperation.

2.2.2 Other Examples of Information Sharing across Organizational Boundaries

While information sharing to support counterterrorism is clearly a significant motivator to information sharing, it is by not means the only motivator. Information sharing furthers other government functions and commercial interests.

Maritime Domain Awareness is a broad imperative directed at achieving effective understanding of *anything* associated with the global maritime domain that could potentially impact the security, safety, economy, or environment of the United States. The scope is obviously very broad in terms of the range of information to be collected, the geographic scope of that information, and the organizations, both government and civilian, that collect that information and/or make use of it. The types of information of interest can include:

- Status of vessels, cargo, and crews
- Maritime areas of interest, including sea lanes or oceanic regions
- Ports, waterways, and their facilities
- Environment, including weather, currents, and natural resources
- Maritime critical infrastructure, including undersea fiber-optic cables and pipelines
- Threats and activities, including illegal migration, drug smuggling, and inherently dangerous activities such as offshore drilling
- Operational information about friendly forces, including those of allies, operating in the maritime domain
- Financial transactions, including illegal money trails and hidden vessel or cargo ownership

Commercial interests, particularly shippers, are interested in knowing the status of their vessels, crew, and cargo. Government organizations, such as the U.S. Navy and the U.S. Coast Guard, are interested in knowing the situation not only on the high seas, but also along the littoral and in ports. Organizations with anti-terrorism missions are interested in being able to identify potential threats well before they near U.S. interests, whether in the homeland or internationally (recall the incident of the attack on the *USS Cole*). The proposed solution to collecting, analyzing, and disseminating this information calls for not only promulgating standards for information sharing, but also establishing a "network-centric, near-real-time virtual information grid that can be shared, at appropriate security levels, by federal, state, local and international agencies with maritime responsibilities."[4]

The kinds of cross-boundary collaboration and information sharing highlighted at the federal level are also occurring among state and local jurisdictions as well as at the regional level (Federowicz, 2006). Consider the following examples of state and regional information sharing:

- The Naval Criminal Investigative Service (NCIS) has launched the Law Enforcement Information Exchange (LinX) initiative, a project to enhance information sharing among federal, state, and local law enforcement in seven regions of strategic importance to the Department of the Navy. LinX provides participating law enforcement agencies with secure access to regional crime and incident data. In the National Capital Region, LinX allows more than 60 state and local police agencies to share mug shots and crime reports.
- In Washington State, the Justice Information Data Exchange was designed to share information about traffic and collision reports with the state Department of Transportation, the Department of Licensing, and the Administrative Office of the Courts.
- In Massachusetts, a consortium with over a hundred participating agencies shares healthcare data through the Massachusetts Simplifying Healthcare Among Regional Entities (MA-SHARE).
- In South Carolina, the Information Sharing and Analysis Center (ISAC) is used to disseminate security-related information to security managers across the state. It is tied into the National ISAC, which includes the Federal Bureau of Investigation, the Department of Justice, and the Department of Homeland Security (DHS).
- In California, the Integrated Nonfiler Compliance system collects data from a variety of federal, state, county, and local sources, including banks, the Internal Revenue Service (IRS), and state licensing boards, among others, to identify individuals who should have filed income taxes and estimates the amount of taxes they may owe.
- In the Washington, D.C. region, the State of Maryland, the Commonwealth of Virginia, and the District of Columbia have established a partnership known as the Capital Wireless Integrated Network to develop an interoperable first responder data communication and information sharing network.[5]
- BioSense is an initiative of the Centers for Disease Control and Prevention. Initially intended as an intergovernmental data collection and analysis system for early detection of bioterror attacks, BioSense and its mission have subsequently been expanded to detect or confirm naturally occurring outbreaks. The system now collects and analyzes data from hospitals, clinics, laboratories, and public health organizations in the government and private sectors.
- In the Washington, D.C. metropolitan region, a collaborative effort among 19 local law enforcement agencies in Maryland, Virginia, and the District

of Columbia led to the development of a well-regarded system to share data about pawn transactions.

These examples of information sharing across organizational, jurisdictional, and functional boundaries, not only among military forces and government agencies but also between government agencies and the private sector, are just a sampling of many such initiatives. In all these cases, technology is the critical enabler. However, it is also true that the technical dimension is often the easy part. Building trust, agreeing on common goals and objectives, and developing the mechanism, processes, and governance structures to collaborate are often more challenging but certainly no less critical than agreeing on common standards and developing the means to collect and share information.

2.3 Enabling Conditions

2.3.1 *Information Technologies Environment*

Information technologies—particularly the computer, the Internet, the World Wide Web, and wireless connectivity—have fundamentally revolutionized how and with whom individuals and organizations interact and conduct business. This has rightly been termed the "information revolution."

The rapid growth and pervasive use of information technologies throughout American society fuels the potential of network-centric operations. Unlike other technologies, in which the military has been both the primary investor and the primary customer, information technologies are driven by the commercial marketplace. Defense and other government agencies leverage industry's large investment in mature and emerging technologies.

The past decade has witnessed the continually falling costs of computing, information storage, and bandwidth; the growth in wireless technologies; and the explosive expansion of the Internet and the World Wide Web. The convergence of these trends has led to omnipresent information technologies, not just in industry, but throughout the economy and not just in wealthy, industrialized nations, but globally—and the users include terrorists

As of August 2003, more than 416 million people around the world had access to the Internet from a personal computer at home. According to data provided by Internet World Stats,[6] an online publisher of world Internet usage and population statistics, by December 2007, worldwide Internet penetration had grown to approximately 1.3 billion people, or a worldwide penetration rate of 20%. Internet usage in North America at the end of 2007 was approximately 238 million users, equivalent to more than 71% of the population. Thus, from 2000 to 2007, usage grew 265.6% worldwide and more than 120% in North America. The rest of the world is rapidly catching up.

2.3.2 Improved Performance and Reduced Costs Drive Information Technologies

In 1965, Gordon Moore of Intel first observed that the number of transistors on a chip was doubling approximately every 12 months and predicted that this trend would continue. This has become known as Moore's law. In fact, this rate of growth continued through the late 1970s; since then the number has doubled every 18 months (Figure 2.1). Continuing miniaturization has translated into constantly improved performance of semiconductor chips, while making them less expensive. As a result, the price of computers has continued to decline and information technology continues to proliferate. Today, embedded microprocessors are ubiquitous, found in printers and copiers, telecommunication devices and manufacturing equipment, and in consumer products such as automobiles and home appliances.

As with semiconductors, the trend in information storage technologies is toward increased performance at lower cost. The shrinking cost of storing and retrieving information has led to significant growth in the amount and types of content stored. Businesses and consumers increasingly create and store digital images and audio and

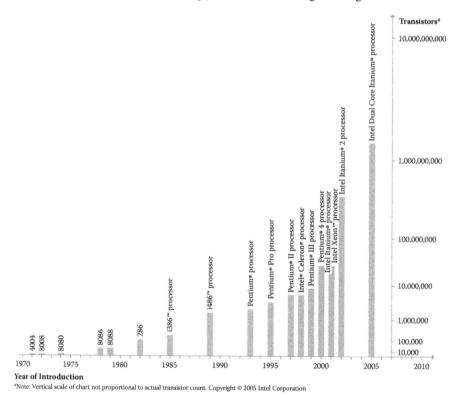

Figure 2.1   Moore's law microprocessor chart.

video presentations. At the same time, they want immediate access, which leads to greater emphasis on online access versus archiving. Moreover, as more data and richer types of data are stored, the need to protect that data becomes more critical. No longer are backup and recovery acceptable; users insist on instant backup and availability.

2.3.3 Global Information Infrastructure

The Internet[7] had its beginning in the DoD Advanced Research Projects Agency Network (ARPANET)[8] research program. Since then, it has become the massive global information infrastructure that allows any computer to communicate with any other computer as long as both are connected to the Internet. It has revolutionized how individuals and businesses broadcast and access information, as well as how and with whom they interact and collaborate. While the widespread commercial use of the Internet is a fairly recent phenomenon, the Internet concept, technologies, and processes evolved and matured over a long period of experimentation and development. Barry Leiner and other architects of the Internet (Leiner et al., 2003) describe the transition from the initial concepts of the Internet envisioned by a small group of researchers to the commercial and social backbone that it has become:

> The Internet has changed much in the two decades since it came into existence. It was conceived in the era of time-sharing, but has survived into the era of personal computers, client-server, and peer-to-peer computing, and the network computer. It was designed before local area networks (LANs) existed, but has accommodated that new network technology, as well as the more recent Asynchronous Transfer Mode (ATM) and frame switched services. It was envisioned as supporting a range of functions from file sharing and remote login to resource sharing and collaboration, and has spawned electronic mail and more recently the World Wide Web. But most important, it started as the creation of a small band of dedicated researchers, and has grown to be a commercial success with billions of dollars of annual investment.
>
> The central concept underlying the Internet was "open architecture networking": that is, that multiple, independent, and heterogeneous networks would interoperate without any central controlling mechanism. Rather than specifying the design of each network, these different networks could be made to work with each other by following a common suite of protocols. That protocol suite came to be known as the Transmission Control Protocol/Internet Protocol (TCP/IP).

As important to the evolution of the Internet as the protocols themselves were the role of documentation and the evolution of the community process. Writing in "A Brief History of the Internet," Leiner et al. (2003) point out that "A key to the rapid growth of the Internet has been the free and open access to the basic documents, especially the specifications of the protocols." Since its origins in a collaborative community of university researchers, the Internet has relied on a community process. Today, the Internet Engineering Task Force (IETF), the principal body responsible for the development of new Internet standard specifications, is a loosely self-organized group of network designers, operators, vendors, and researchers who contribute to the engineering and evolution of Internet technologies. The IETF is open to any interested individual.

As the Internet has grown, so has its value. Robert Metcalf, founder of 3Com Corporation and a major designer of the Ethernet (a family of local area network products) is credited with the development of Metcalf's law, which states that the value of a network increases exponentially with the number of nodes; that is, the usefulness of a network is proportional to the square of the number of users. In essence, the more users who are connected, the more useful the network becomes. The original formulation was intended to convince early Ethernet adopters to establish local area networks that were large enough to exhibit this network effect. That notion was subsequently named Metcalf's law in a paper by Gilder (1993).

> In this era of networking, he [Metcalf] is the author of what I will call Metcalf's law of the telecosm, showing the magic of interconnections: connect any number, "n," of machines—whether computers, phones or even cars—and you get "n" squared potential value. Think of phones without networks or cars without roads. Conversely, imagine the benefits of linking up tens of millions of computers and sense the exponential power of the telecosm.

More recently, David Reed has proposed that the value of the network lies not merely in its ability to connect peers, as in Metcalf's law, but even more in its ability to allow users to form and maintain groups, such as auction sites and chat rooms. Reed's law (Reed, 2001), as it has become known, suggests that such group-forming networks have a value that grows exponentially, in proportion to 2^n. According to Reed's law, the value of group-forming dominates the earlier broadcast (one-to-many) and peer-to-peer (one-to-one) transactional network.

The term "World Wide Web" (WWW) or the "Web" refers to all the information sources that can be accessed using a Web browser. It is a distributed, hypertext-based information system that provides universal access to a wide range of material, including text, images, graphics, and sound. The technologies underlying the Web were originally developed as a means for sharing information about high-energy physics among scientists worldwide. The Centre Européenne pour la Recherche Nucléaire (CERN), the European Laboratory for Particle Physics, developed the Web syntax, specifically the Hypertext Transport Protocol (HTTP), used by WWW servers, and the Hypertext Markup Language (HTML), a language that governs the creation of Web documents so that they can be read by Web browsers. In 1992, the National Center for Supercomputing Applications (NCSA) introduced NCSA Mosaic™, the first readily available graphical Web browser.

The Web, like the Internet, has no central management. Anyone can publish and there is no central dependence on a single server. Also like the Internet, Web technologies are advanced by a cooperative community process, the World Wide Web Consortium (W3C).

The growth and penetration of the Web have been spectacular. What started out as a way to share scientific papers had, within a decade, exploded to penetrate the retail, education, entertainment, and news sectors.

Businesses and other organizations have established private networks using Internet and Web technologies to allow users to find, use, and share documents and Web pages. In some large companies, intranets have become the primary means for employees to obtain and share work-related documents, share knowledge, collaborate on designs, access e-learning, and learn about company news. Intranets use traditional Internet protocols to transfer data. They usually reside behind firewalls for security, and are not limited by physical location: anyone anywhere in the world can be on the same intranet. Intranets also link users to the outside Internet and, with the proper security in place, may use public networks to transfer data.

Like intranets, extranets are private networks that use Internet protocols. The difference is that they are designed to provide a safe way to allow transactional business-to-business activities between an organization and its suppliers, partners, or customers. For example, the automotive industry uses extranets to cut down on its redundant ordering processes and keep suppliers up-to-date on parts and design changes, thereby shortening response times to suppliers' problems and questions. Suppliers can receive proposals, submit bids, provide documents, or collect payments through an extranet site. Because an extranet has restricted access, it can be connected directly to a company's internal systems.

2.3.4 Mobile Society, Wireless Lifestyle

The emerging wireless lifestyle goes well beyond the use of cellular phones and pagers to a fundamental extension of the Internet from a fixed, wired environment to one that is wireless, allowing access to information anywhere and anytime. Today, cafés, restaurants, hotels, malls, and other businesses offer "hotspots"—wireless extensions of existing high-speed broadband Internet accounts. Any wireless fidelity (WiFi)[9]-enabled device can access the Internet while operating within range of the wireless extension. In addition to these "fixed" wireless sites, consumers can make use of mobile wireless—operating aboard motorized, moving vehicles—and portable wireless. Applications include not only telecommunications, but also entertainment, interpersonal communications, multimedia messaging, mobile payments, and location-based information services.

A number of key technologies underlie this significant trend. First, pocket-sized wireless communication devices are proliferating. These integrate voice and data, and support multiple functions, including voice communications, e-mail and text messaging, and Internet access, as well as many other capabilities normally associated with fixed computers. Continuing improvements are leading to devices of smaller size, lower weight, and extended battery life.

Second, the wireless communications networks are improving in connectivity, capacity, and security. Examples of wireless communication technologies include General Packet Radio Service (GPRS), a packet-based wireless communication service that provides continuous connection to the Internet for mobile phone and computer users, and Wireless Application Protocol (WAP), a set of communication protocols to standardize how wireless devices can be used for Internet access.

Location-aware computing is another wireless trend that leverages the position location services of the Global Positioning System (GPS). It promises to provide individuals and businesses with localized services, such a local maps, as well as the ability to locate and track key people and assets.

2.3.5 Increased Reliance, Increased Risk

The explosive growth of the Internet, intranets, and extranets reflects their increasing importance for government and business operations as well as for individual usage. As the Internet has become more critical to achieving business goals—reducing operating costs, increasing organizational collaboration, increasing sales—it has become more tightly coupled to the organization's business practices. In effect, what has emerged is a rich set of network-enabled business processes, such as Enterprise Resource Planning (ERP) systems, network operations, and electronic commerce operations, that inherently depend on the availability of the network. Increased dependence on the network means that any disruption of the network can paralyze an organization's fundamental business processes.

Disruptions to the network come from multiple sources, both accidents and deliberate attacks. In 1999, for example, a backhoe inadvertently cut a fiber cable, resulting in nationwide disruptions of communications. Traffic had to be rerouted around the cut cable, causing congestion in the other parts of the network. Traffic between the East and West Coasts was as much as 50 times slower, and some affected companies had to close down their operations temporarily (*New York Times*, 30 September 1999). In 2007, a fire started by a homeless man disrupted service between Boston and New York on the experimental Internet 2 network (Gaffin, 2007).

But the most publicized disruptions have resulted from malicious attacks. During one week in 2000, Distributed Denial-of-Service (DDS) attacks brought down many well-known Internet sites, including Yahoo!, Buy.com, eBay, Amazon, Datek, E*Trade, and Cable News Network (CNN). In such a DDS attack, hackers remotely install software called "bots" on hundreds or thousands of machines on the Internet and use these machines to launch a coordinated attack on a targeted system. The flood of incoming Internet Protocol (IP) packets can force the targeted site to shut down, thereby denying service to legitimate users.

In July 2001, Code Red, a self-propagating worm,[10] began to exploit vulnerabilities in Microsoft's Internet Information Server. Later that same month, Code Red II was launched, and shortly after September 11, 2001, NIMDA ("admin" written backward) was detected. NIMDA started in the United States and, within an hour, had infected 86,000 computers. By the next day it had spread to Europe and Asia, where it was disrupting Internet traffic. (It is worth noting that these worms exploited well-recognized vulnerabilities for which patches had been available for over a year.)

In October 2002, an attack on the 13 root servers[11] that manage worldwide Internet service brought down seven of those servers and caused two others to fail intermittently. Interestingly enough, while the attack was massive, it actually did little to disrupt the Internet, largely because of the heavy reliance on the local servers used to handle requests initially. It is only when the local server cannot resolve an address that a request is forwarded to one of the root servers.

These and other such examples highlight the vulnerability of the Internet to asymmetric threats. Relatively few resources—in these cases, an unwitting backhoe operator or a few hackers armed with readily available scripts—are needed to attack a network and cause significant impact. This, in turn, increases the attractiveness of a cyber attack on the network for adversaries who are motivated by financial gain, ideology, revenge, or even simple mischief.

The complexity of the network and, in particular, of software increases vulnerabilities that can be exploited. At the same time, the very complexity of the network and its "loose coupling" and "scale-free"[12] aspects serves to insulate it to some degree. Here, "scale-free" refers to situations in which most of the nodes of a network will be loosely connected, while a minority will be very richly connected.

Regardless of the source or motivation, a disruption can have potentially massive economic impacts, and those impacts can be global. Damage can include lost sales, the need to replace hardware, hampered productivity, and cost to reputation. In addition, there are increasing costs associated with preventing, detecting, and repairing attacks.

2.4 Institutional Trends: Enterprisewide, Top-Down Perspective

Investments in information technologies enable increased interconnectivity and increased interoperability. The potential of this interconnectivity and interoperability has fueled fundamental changes in how organizations expect to conduct their operations. Yet, at the same time, government agencies and corporations are hampered by the existing inventory of systems—often referred to as legacy systems. These systems may have been developed for specific parts of the enterprise, are often tailored to local needs, and can be difficult to adapt to the new business models. Rather than enabling interconnectivity and interoperability, these legacy systems constitute barriers. For example, General Motors (GM) discovered that the company had more than a dozen SAP AG[13] systems deployed globally, each of which was a different release and was used in slightly different ways. The resulting incompatibilities slowed GM's progress toward its goal of having a common mobile system (CIO Staff, 2004).

Until recently, enterprises made their investment decisions locally to support a particular function or business unit, resulting in separate, often "stovepiped" systems that mirrored the "stovepiped" organizational units they supported. This did not present a problem as long as the supported units continued to operate independently. However, when organizations wanted to realign their organizations and processes, improve performance, reduce costs, or develop new capabilities, they found that they were hampered by gaps, incompatibilities, and/or systems that did not allow ready adaptability.

The response of both the government and commercial sectors has been to take an enterprisewide and top-down perspective toward defining the capabilities that the enterprise needs to implement. By *enterprisewide* we mean all the organizational elements and associated resources—people, processes, and information—that work together to achieve a common mission. By *top-down* we mean that the process is driven by the enterprise's vision, mission, and strategic agenda and uses these to develop and manage its investments, including those associated with its information infrastructure. The investment portfolio, in turn, defines the projects that will be implemented. Contrast this with the traditional approach to investments, in which each organizational unit defines its own needs and structures its own projects more or less independently of others in the enterprise.

2.4.1 Clinger-Cohen Act of 1996

The Clinger-Cohen Act of 1996[14] mandates this enterprisewide and top-down perspective for U.S. federal agencies. It explicitly requires that investments in information technologies be linked to the accomplishment of the agency's mission and goals. In addition, it directs that agencies link their IT investment process to their capital planning process. A key provision of the Act was establishment of the role of Chief Information Officer (CIO) in each agency—a position that had emerged in the early 1980s in industry. Agency CIOs report directly to the agency executive and have information management as their primary portfolio. This has had several key effects. First, it elevates the position to an executive level and establishes an advisory relationship between the CIO and the chief executive officer. It also emphasizes the strategic linkage between investments in IT and the agency's mission. This parallels the evolution of the role of CIO in the commercial world from a purely technical support function—in effect, a service provider—to an executive, strategic position.

The Clinger-Cohen Act instituted various other important measures. Among them are business process reengineering as a precursor to information technology (IT) investment, prioritization of IT initiatives using analysis of alternatives and return-on-investment techniques, and development of integrated agency-wide IT architectures.[15] Now termed "enterprise architectures," they are intended to link the business's IT strategy to the organization's mission, strategy, and processes.

2.4.2 Enterprise Architectures

Enterprise architecture is expected to play a central role in the enterprise life cycle (Figure 2.2) (Cady, 2003). U.S. federal agencies face a renewed mandate to develop agency-wide enterprise architectures to improve planning and better understand the impact of technology investments on their overall agency operations. The U.S. Office of Management and Budget (OMB) currently requires agencies to define their enterprise architectures within the scope defined by the OMB's Federal Enterprise Architecture (FEA) Office before agency programs can receive funding. The OMB expects that, by implementing the FEA approach, agencies will achieve greater internal efficiencies, streamline business operations, and improve interagency collaboration.

An enterprise architecture is expected to describe all aspects of an organization—its mission, organizational structure, business processes, and information exchanges—and relate them to the enabling information resources (see Figure 2.3). Its goal is to align all business functions and operations with the mission and to identify the changes necessary to carry out the agency's strategic plan. A sequencing plan describes the order in which these changes will be made.

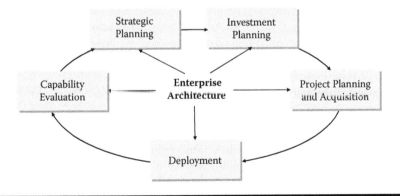

Figure 2.2 Enterprise life cycle. (Adapted from Cady, A. 2003. Technologies for Enterprise Modernization, The MITRE Corporation, McLean, VA, 30 September 2003).

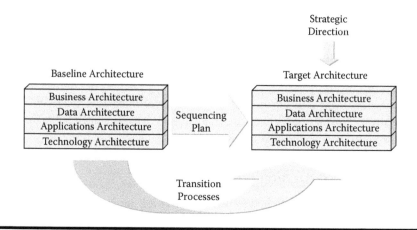

Figure 2.3   Elements of an enterprise architecture.

Although most discussions of enterprise architecture are based on the organization as the enterprise, an enterprise can—and often does—extend beyond a single organization. The enterprise can be a functional area that involves multiple organizations, as in the case of supply chain partners, or can even involve several functional areas. The architecture of an extended enterprise is envisioned as a potent tool that would enable collaboration among the participants of such a cross-organizational enterprise.

2.4.3 Changes to the DoD Requirements Process

In the U.S. DoD, recent changes in key management processes also emphasize this new enterprisewide and top-down perspective. Most prominent is the revised requirements generation process. Traditionally, the DoD implemented requirements from the bottom up, with each military Service and agency defining its own needs and proposing its own programs. Frequently, the resulting systems could only be integrated after they were built, and then often with considerable difficulty and cost. This bottom-up approach has been viewed as overemphasizing the interests and needs of the individual military Services at the expense of joint warfighting needs. It can also yield potentially duplicative capabilities—different approaches to achieving the same end.

In response, the DoD has promulgated a radically different process (Figure 2.4). The Joint Capabilities Integration and Development System (JCIDS)[16] is deliberately top-down and builds on strategic plans and warfighting visions. Through a series of structured analyses, it identifies and describes necessary capabilities and gaps. These capabilities and gaps, in turn, form the basis for defining the investments to be made. By bringing more organizations into the early phases of the

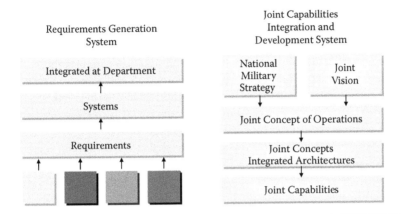

Figure 2.4 Transformation from requirement generation system to Joint Capabilities Integration and Development System (JCIDS).

process, JCIDS is intended to improve coordination and provide a broad-based review of capability proposals.

The JCIDS defines a series of documents that are linked to the acquisition process. These documents become increasingly specific over the life of a project, starting with identification of a capability gap, continuing by identifying the attributes of a proposed system, and progressing to identifying production attributes for a single increment of a program.

Concurrently, the DoD revised its acquisition policy.[17] A key dimension of the new acquisition policy is that it makes evolutionary acquisition the preferred strategy for rapid acquisition of mature technologies (Figure 2.5). In so doing, it de-emphasizes the traditional single-step approach—sometimes referred to as the "big bang" approach—and institutionalizes an acquisition strategy that emphasizes incremental fielding of capabilities as they mature.

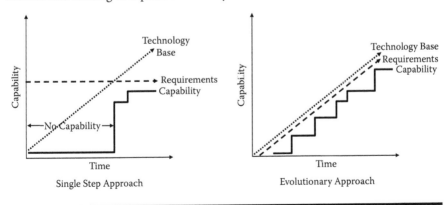

Figure 2.5 Comparison of single-step and evolutionary acquisition approaches.

2.5 Implications for Systems and Programs

How do these operational, technical, and institutional trends affect the systems that are and will be developed? They have several significant implications.

First, there is a trend toward increased program *scale and scope* as single acquisition programs encompass what in the past would have been separate acquisition efforts. The U.S. Army's Future Combat Systems (FCS) is one such example. In this case, a single acquisition program is responsible for acquiring and integrating a range of manned and unmanned platforms as well as the communications, command and control, sensors, and processing capabilities that tie them together.

Commercial and government enterprises are also seeking to integrate separate, often isolated, operations, processes, and information. In so doing, they are taking an organization-wide perspective on how they organize and operate. Decisions about investments in individual information technologies, previously made locally, are now being made at the enterprise level.

A related trend is the *convergence of previously independent systems*. It is not only newly initiated programs that are being affected. Programs that were previously separately managed are being organized into cooperative efforts. For example, the Global Command and Control System (GCCS) has had several variants, each focused on meeting the particular needs of the individual funding organizations. They sought to unite these separate efforts (Joint, Army, Maritime, and Air Force) in a common development and engineering effort called the Net-Enabled Command Capability (NECC). While the NECC allowed for extensions to meet unique service needs, the intent was to emphasize a common core capability. Development was to be allocated to the different program offices based on their specific areas of expertise.

Similarly, the U.S. Missile Defense Agency was established to oversee and direct the development of a layered Ballistic Missile Defense System (BMDS) that will combine several programs into one system capable of destroying an enemy missile from shortly after launch to shortly before impact on the intended target. These previously separate programs, now termed "elements," include land-based systems such as the Patriot Advanced Capability-3 and the Terminal High Altitude Air Defense,[18] sea-based systems such as the Aegis Ballistic Missile Defense, and space-based systems such as the Space Tracking and Surveillance System.

The combination of increased scale and scope and convergence of previously separated systems translates into systems that *cross traditional boundaries*. These boundaries can be organizational, as in the case of the NECC, which had been initially organized along service interests, or they can be based on function or discipline. As in the example of the Army's FCS, we would expect to see systems that span the areas of operations, logistics, and training—domains that until now were supported by separate systems and managed by separate communities. The same holds true across discipline boundaries, such as the various intelligence disciplines (e.g., signals intelligence or imagery intelligence). Crossing such traditional boundaries is difficult and not always successful. It is worthwhile noting that in 2009, Congress, in effect, directed the termination of the NECC program, citing a lack of coordination between the services on how to proceed.

Information technologies will continue to be at the core of these emerging, large-scale systems as developers seek to leverage commercial technologies and common, often commercial, standards. As this occurs, there will be continued growth in *integration* and a commensurate decline in custom developments. The integration challenge will continue to increase as the efforts focus on integration of heterogeneous components. Not only do we expect the components to be diverse, but the development activities will also be distributed across multiple, often physically dispersed, activities that may or may not report within a common organizational structure.

In addition to the above primarily structural and organizational considerations, additional implications touch on the *requirements for the systems* themselves. These systems must accommodate rapidly evolving needs, organizational patterns, and enabling technologies. We will not always be able to state, with any reasonable precision and certitude, a set of required attributes that can be expected to remain constant over the course of a lengthy development effort. Instead, we anticipate that the needs will evolve in parallel with and often in response to the evolution of the systems themselves.

The concept of "unlikely partners" applies here. Traditionally associated with the world of mergers and acquisitions, this concept connotes alliances between organizations that have either been direct competitors or that, despite substantially different orientations, approaches, and interests, need to come together to work toward some common objectives. This concept is paralleled in the DoD, notably in coalition and interagency operations, as illustrated by operations in Afghanistan. In Operation Enduring Freedom, we have seen operations that integrated Army and Marine Corps forces, Special Operations Forces (SOF), other government agencies (notably the Central Intelligence Agency), and indigenous forces. No architecture or use case could have anticipated this particular set of partners, and certainly no information exchange requirement could have predicted the specific pattern and content of their interchanges.

The same notion of "unlikely partners" applies not only during operations, but also during development. The broader the scope of the system, the more diverse the set of stakeholders who may come to the table with different administrative, cultural, technological, and organizational perspectives and different objectives they wish to pursue.

Unlikely partners can, and do, occur among government agencies and between federal, state, and local governments. They can also occur in business. Consider the business and technical collaboration agreement announced in 2006 between arch competitors Novell and Microsoft to make their products, the Linux platform and open source software, and Microsoft's proprietary Windows®, work better together. They agreed to collaborate on the development of specific technologies, to promote each other's products, and to create a joint research facility. In addition, they provided each other's customers with assurance against patent infringement claims, giving their customers confidence that the technologies they use are compliant with the patents of both companies. In addition, Microsoft agreed not to enforce its patents with individual, non-commercial Linux developers. A joint letter to the Open Source Community from Novell and Microsoft stated that "Today's announcement of the collaboration between Microsoft and Novell marks the beginning of a new era: Microsoft is coming to terms with Linux." Moreover, it goes on to say:

> Why is Microsoft doing this? Because they recognize that customers today are deploying mixed source solutions—Windows and Linux— and they want these solutions to work well together. This will help Microsoft by making it easier for Linux customers to deploy Windows in their Linux environments. Microsoft is committing significant resources to promote joint Windows-Linux solutions. This is all about co-existence and giving customers greater choice.

As an interesting aside, while this agreement was being formulated, Novell continued action on a suit, filed in 2004, that alleged that Microsoft had used anticompetitive practices against an earlier Novell product line, the WordPerfect office software suite.

Finally, the emerging systems are expected to be increasingly *complex*. The side effect of having systems that accommodate multiple communities and interests, and are themselves evolving, is that the system behavior will not always be predictable, but instead will emerge from the interaction of the components.

2.6 A Look Ahead

This chapter briefly sketched a view of the current situation and the near future, primarily using U.S. defense as the context: rapidly evolving, large-scale, richly interconnected systems intended to bridge traditional boundaries. These systems are not merely scaled-up versions of the systems that were developed in the latter half of the twentieth century, but instead represent a significant departure. The practice of systems engineering has evolved over the past half century and will inevitably continue to evolve to meet the requirements imposed by this new class of systems. The traditional processes and practices must be reexamined for their efficacy and suitability in this new, more challenging systems engineering environment.

Endnotes

1. Available online at www.defenselink.mil/speeches/speech.aspx?speechid=488 (accessed 27 February 2008).
2. The transcript of this briefing is quoted in Cordesman (2003) and is also available online at: http://www.globalsecurity.org/wmd/library/news/iraq/2003/iraq-030322-centcom03.htm (accessed 27 February 2008).
3. "Network-centric operations" are defined as military operations that are enabled by the networking of the force. When these operations take place in the context of warfare, the term "network-centric warfare" applies (DoD, 2001a, b).
4. *The National Plan to Achieve Maritime Domain Awareness*, published in October 2005, establishes a set of near- and long-term actions and assigns responsibility for them (DHS, 2005).
5. In November 1998, traffic on the Woodrow Wilson Bridge spanning the Potomac River was brought to a halt for several hours as a man threatened to jump from the bridge. First responders from the State of Maryland, the Commonwealth of Virginia, and the District of Columbia found that lack of interoperability hampered their ability to coordinate and respond to incidents such as these.
6. Available online at www.internetworldstats.com (accessed 27 February 2008).
7. In 1995, the Federal Networking Council passed a resolution defining the term "Internet": RESOLUTION: The Federal Networking Council (FNC) agrees that the following language reflects our definition of the term "Internet". "Internet" refers to the global information system that—(i) is logically linked together by a globally unique address space based on the Internet Protocol (IP) or its subsequent extensions/follow-ons; (ii) is able to support communications using the Transmission Control Protocol/Internet Protocol (TCP/IP) suite or its subsequent extensions/follow-ons, and/or other IP-compatible protocols; and (iii) provides, uses or makes accessible, either publicly or privately, high level services layered on the communications and related infrastructure described herein.
8. The ARPANET, a computer communications network, was developed under the sponsorship of the Department of Defense, Advanced Research Projects Agency (ARPA), as it was then known.
9. WiFi stands for and also refers to wireless Internet being 802.11 enabled.
10. A worm is a computer virus that is designed to copy itself and thereby spread across the network, most often by e-mail, but also via other routes. As the worm spreads, it creates additional traffic that clogs servers. Some worms can also damage computer systems.
11. The Domain Name Service (DNS) is an Internet service that translates Internet address names into their corresponding numeric IP addresses. There are currently 13 DNS root servers, managed by different organizations, spread around the world. Initial requests are normally routed to a local server. If the address cannot be resolved locally, then it is passed to the root server.
12. Random networks are those in which most nodes have approximately the same number of links. By contrast, scale-free networks obey a power law distribution. In the latter case, while most nodes have only a few links, a few nodes have a very large number of links. For a more detailed discussion of scale-free networks and the implications for network topology, see Barabasi (2002).

13. SAP AG is a German company that develops a wide range of enterprise software applications.
14. The fiscal year 1997 Omnibus Consolidated Appropriations Act, Public Law 104-208, renamed both Division D (The Federal Acquisition Reform Act) and E (the Information Technology Reform Act) of the 1996 DoD Authorization Act, Public Law 104-106, as the "Clinger-Cohen Act of 1996."
15. Clinger-Cohen defines information technology architecture as "an integrated framework for evolving and maintaining existing information technology and acquiring new information technology to achieve the agency's strategic goals and information resources management goals."
16. Formalized in Chairman Joint Chiefs of Staff (CJCS) Instruction *CJCS 3170.01E*, dated 11 May 2005, which provides a top-level description and organizational responsibilities. The companion CJCS Manual (CJCSM) *CJCSM 3170.01B* describes the JCIDS analysis process, defines performance attributes and key performance parameters, describes the validation and approval process, and defines document content.
17. DoD Directive 5000.1 was revised to retain principles and emphasize innovation and flexibility. DoD Instruction 5000.2 was rewritten to focus on required outcomes and statutory requirements and less on regulatory requirements. DoD Regulation 5000.2-R was canceled. It became a "guide" that provides expectations, best practices, and lessons learned.
18. Previously, Theater High Altitude Air Defense.

CONCEPTS AND FRAMEWORKS

Chapter 3

Mega-System Concepts

The preceding chapter (Chapter 2) discussed the forces—operational, technical, and institutional—moving us toward the implementation of large-scale, richly interconnected systems. These systems have been referred to as "systems of systems," "families of systems," "enterprises," and other such terms connoting that they are composed of elements that are purposeful systems in their own right. Because these terms mean different things to different communities, the term "mega-systems" has been chosen as an umbrella term to refer to this class of systems.

It is important to note that there is no community-wide agreement on terminology and, consequently, no common lexicon. The difference in terminology reflects, to a large degree, different perspectives on the same phenomenon.

Mega-systems are characterized not only by scale (larger versus smaller), complex behavior, and the fact that they cross traditional boundaries, but also by "nestedness." Just as they are composed of multiple separate systems, they themselves constitute part of one or more even larger mega-systems. Moreover, to achieve their goals and objectives, they must interact with external but related systems.

3.1 What Is a System?

Professional organizations, standards bodies, and the professional literature have different definitions of the term "system." Examples of such definitions include the following:

- "A set or arrangement of things so related or connected as to form a unity or organic whole" (Webster, 1996).
- "A set of interconnected elements that achieve a given objective through the performance of a specified function" (IEEE, 1997).
- "The top element of the system architecture, specification tree, or system breakdown structure that is comprised of [sic] one or more products and associated life-cycle processes and their products and services" (IEEE, 1997).
- "An interdependent group of people, objects, and procedures constituted to achieve defined objectives or some operational role by performing specified functions. A complete system includes all the associated equipment, facilities, materiel, computer programs, firmware, technical documentation, services, and personnel required for operations and support to the degree necessary for self-sufficient use in its intended environment" (IEEE, 1997).
- "A specific grouping of subsystems, components, or elements designed and integrated to perform a military function" (OSAF, 2000).
- "1. The organization of hardware, software, material, facilities, personnel, data, and services needed to perform a designated function with specified results, such as the gathering of specified data, its processing, and delivery to users. 2. A combination of two or more interrelated pieces of equipment (or sets) arranged in a functional; package to perform an operational function or to satisfy a requirement" (DAU, 2005).

The term "system" can also refer to social and economic systems. Thus, Russell Ackoff (1994) defines a social system as a "whole that cannot be divided into independent parts". He goes on to state that "the performance of a system obviously depends on the performance of its parts, but as important, if not the most important aspect of a part's performance is how it interacts with other parts to affect the performance of the whole."

While the definitions cited above differ in their perspectives and thus their specifics, they have several characteristics in common:

1. They all agree that a system is composed of elements. These elements, which can be hardware, software, or even "liveware" (people and procedures), have well-defined functions within the system context.
2. The elements are somehow interconnected or integrated.
3. The elements, working together, perform specific functions.
4. Inherent in the concept of a system is the notion of self-sufficiency; that is, the system as a whole achieves a given objective through the performance of its constituent elements.
5. The system has a boundary that distinguishes it from the environment in which it operates. The boundary may be physical, as in the case of a visible

entity, or conceptual. The system may stand alone or it may interact with other systems outside its boundary. In both cases, there is a clear distinction between inside and outside.

It is interesting to note that none of these definitions mentions scale. In fact, a system can range from something quite small physically, such a personal computer, to something very large, such as an airplane or an assembly line. It can be a physical construct, a social and organizational entity, or even an organism. It can be engineered—that is, man-made—or it can exist in the natural world.

3.1.1 Systems as Components of Larger Systems

Just as systems are made up of components—subsystems—so, too, systems can interact with related external systems to provide a desired capability. In effect, there are three tiers: (1) the system, (2) the components of the systems (its parts), and (3) the larger system (the "containing system" in Ackoff's terms) of which it is a part. For example, an aircraft navigation system consists of various components and is itself part of a larger system, the aircraft system. In turn, the aircraft system interacts with other systems, such as the air traffic control system, to form the overall air transport system, which in turn represents one part of the national and global transportation network.

From the perspective of a higher tiered system, the parts are components. From the perspective of a lower tiered system, the larger whole is the containing system—the environment or context. This is the notion of *nestedness* (Figure 3.1).

To complicate matters further, a system can simultaneously be a component of more than one "containing system." Thus, a weapon system can be a component of that military service's mega-system while at the same time being a component of the joint mega-system specific to a particular theater of operations. Similarly, an organization's information system can serve the company's internal operations while simultaneously enabling information to be shared externally with its suppliers and other strategic partners.

Another example comes from the DoD. Writing in 1996, Admiral William A. Owens, then Vice Chairman of the Joint Chiefs of Staff, described the vision of an emerging U.S. military system-of-systems that, in his words, "...is at the heart of the American revolution in military affairs (RMA). It embodies a new appreciation of joint military operations, for the system-of-systems depends ultimately on contributions from all the military Services, a common appreciation of what we are building, and a common military doctrine" (Owens, 1996).

This vision, which would result in a "qualitative jump in our ability to use military force effectively," entails weaving together three separate "technology paths," each of which is very large and complex in its own right:

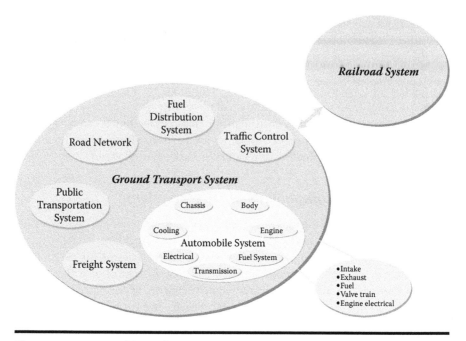

Figure 3.1 Systems hierarchy.

1. Intelligence, surveillance, and reconnaissance technologies, including sensors and reporting technologies as well as the means to keep track of our own forces, to dramatically expand the capability to maintain real-time, all-weather awareness of what is occurring throughout a large geographical area
2. Command, control, communications, and computer applications and intelligence processing technologies to translate awareness of what is occurring on the battlefield into understanding and into directives for forces to execute
3. Precision force technologies, including, but not limited to, precision-guided weapons

3.1.2 What Is a "System-of-Systems?"

The term "system-of-systems" is most commonly used to refer to systems made up of elements that are themselves systems. A system-of-systems can, in fact, be divided into its independent parts.

At lower levels in the hierarchy, the system-of-systems can be physically bounded on a single platform (such as the automotive system shown in Figure 3.1). At higher levels it becomes more abstract, consisting of a grouping of independent but interacting systems.

While there is no universally accepted definition, perhaps the best—certainly the most often repeated and cited—characterization of a system-of-systems was provided by Maier (1996). Maier differentiates a system-of-systems from other very large and complex, but monolithic, systems. For Maier, a true system-of-systems has the following characteristics:

1. *Operational independence of the elements.* If the system-of-systems is disassembled into its component systems, the components must be able to usefully operate independently. The system-of-systems is composed of systems that are independent and useful in their own right.
2. *Managerial independence of the elements.* The component systems not only *can* operate independently, they *do* operate independently. The component systems are separately acquired and integrated but maintain a continuing operational existence independent of the system-of-systems.
3. *Evolutionary development.* The system-of-systems does not appear fully formed. Its development and existence is evolutionary with functions and purposes added, removed, and modified with experience.
4. *Emergent behavior.* The system performs functions and carries out purposes that do not reside in any component system. These behaviors are emergent properties of the entire system-of-systems and cannot be localized to any component system. The principal purposes of the system-of-systems are fulfilled by these behaviors.
5. *Geographical distribution.* The geographical extent of the component systems is large. Large is a nebulous and relative concept as communications capabilities increase, but at a minimum it means that the components can readily exchange only information and not substantial quantities of mass or energy.

Maier (1996) also points out that systems can be classified according to the approaches used in managing them. He identifies three categories: directed systems, collaborative systems, and virtual systems. These categories also apply to systems-of-systems:

1. *Directed* systems-of-systems are built and centrally managed to fulfill specific purposes. Maier (1996) cites an integrated air defense system as an example of this class of systems-of-systems because, while its component systems may operate independently, the entire system is usually centrally managed to defend a region against enemy air attack.
2. *Collaborative* systems-of-systems have a central management authority but lack coercive or directive power. As in the Internet, components of such systems voluntarily collaborate to fulfill agreed-upon central purposes.
3. *Virtual* systems-of-systems lack both a central management authority and, indeed, a centrally agreed-upon purpose. As in the example of the World Wide Web, large-scale behavior emerges but it does so through relatively invisible mechanisms.

3.1.3 What Differentiates "System-of-Systems" from Similar Terms?

As writers struggle to define the characteristics of a system-of-systems, new terms are introduced to highlight particular features and dimensions.

A *federation of systems* is a special case of a system-of-systems. Krygiel (1999) defines a federation of systems as one in which the component systems are managed by separate—although collaborating—organizations, each with its own methods, technologies, and schedules. Consequently, the components of a federation of systems are more autonomous, more heterogeneous, and more widely distributed than those that make up a system-of-systems. Thus, the term "federation of systems" focuses on the managerial independence of the systems that comprise the system-of-systems. An example of such a federated system is a modeling and simulation system that uses a common infrastructure to exchange information among individually developed models.

The term *network of systems* focuses on another aspect of Maier's characterization: geographical dispersion. In this case, the component elements are geographically distributed and their interactions are limited to information exchange through the communications network (Matthews et al., 2000) The Internet is one example; a network of distributed sensors is another.

A *family of systems* has a fundamentally different connotation. While it also consists of separate systems, it is more akin to a product line in which the members of the family share certain features for consistency and efficiency while allowing for necessary specialization to accommodate the range of needed capabilities. To that extent, the family of systems consists of independent systems that can be arranged or tailored in various ways to meet specific needs. An example of such a family of systems is the former GCCS. The GCCS Family of Systems (as it was indeed referred to) consisted of a number of service-specific variants, each developed on a common software infrastructure. A separate program office managed the development of each variant. A second example of a family of systems is the Army Future Combat System. This program, under a single Army program manager, was responsible for developing and managing a number of manned ground combat vehicles, unmanned ground and aerial vehicles, and unattended sensors and munitions, many of which share common components for reasons of improved efficiencies and low logistics burdens. Tying these individual systems together is a network of Command, Control, Communications, Computers, and Intelligence, Surveillance, and Reconnaissance (C4ISR) capabilities.[1]

An *enterprise system* is a social system. It is an organization of people and other resources in one or more locations with a common mission that is implemented by automated or semi-automated business processes, associated information exchanges, and supporting technical infrastructure. It can be a single organization or it can include external partners. When it includes external partners, it is often referred to as an "extended enterprise." Corporations, government agencies, and membership organizations are all examples of extended enterprises. From the perspective of the computer industry, an enterprise is both the organization that uses computers and the large-scale, organization-wide network that enables the organization to accomplish its missions.

What is the significance of these different terms?

1. They all address a common concept that is well recognized and, in many cases, deliberately sought after: achievement of a fundamentally new and desirable capability through the cooperative interaction of what may otherwise have been separate systems.
2. More than a decade after Maier's paper, it is clear that while there may be an intuitive understanding and wide recognition of this concept, there is still no commonly accepted terminology.
3. The very existence of multiple terms indicates that different authors have each focused on particular dimensions of the same problem and suggests that no one-dimensional view will be sufficient. Clearly, multiple perspectives are needed instead.

3.2 Mega-Systems

The term "mega-systems" is used as a convenient umbrella term to encompass notions of systems-of-systems, federations of systems, networks of systems, enterprise systems, and other such closely related terms. The term alludes to both the scale of these systems and the fact that they cross traditional boundaries.

3.2.1 What Are Mega-Systems?

"Mega-systems" are the large, complex systems that cross traditional boundaries to provide a level of functionality not achieved by their component elements. This definition encompasses the following salient characteristics:

1. They are *large, man-made systems*. While "large" is clearly a relative term, these systems provide multiple functions, support multiple users, and may be distributed over a wide geographic area. They may support an enterprise or extend across multiple organizations that cooperate in achieving a common mission or objective.
2. They are *complex*. By "complex," we do not mean that they are complicated or difficult to construct, which they often are, or even that they have many component parts, which they often do, but that they exhibit complex behavior,

both internally among their components and as a whole. Senge (1990) and Sterman (2000), both from MIT's Sloan School of Management, also make this distinction, differentiating between detail complexity and dynamic complexity. Detail complexity exists when a system has many components or a problem has many variables. Nevertheless, such complexity is tractable, given the right tools and sufficient resources. Dynamic complexity, on the other hand, is fundamentally different. "When the same action has dramatically different effects in the short run and the long, there is dynamic complexity. When an action has one set of consequences locally and a very different set of consequences in another part of the system, there is dynamic complexity. When obvious interventions produce nonobvious consequences, there is dynamic complexity" (Senge, 1990).

Internally, there are many possible interactions, some of which are predictable and expected, but others that are neither. Changes in the behavior of one element can—and do—have an impact on the behavior of other elements, often in unpredictable ways and under unanticipated conditions. (In medicine, this is known as "side effects"; more generally, this is referred to as "unintended consequences.") The behavior of the mega-system as a whole cannot be inferred simply from knowing the behavior of each of its constituent elements. Rather, complex systems exhibit "emergent behavior": behavior that accrues to the whole and is neither predictable from nor resident in the behavior of its constituent elements. In simpler terms, "the whole is greater than the sum of its parts."[2]

3. They *cross traditional boundaries* and do so intentionally. These boundaries are like fences in that they formalize and, in many cases, limited the interactions between the "inside" and the "outside." They could be functional boundaries, such as intelligence and operations in the military domain or marketing and engineering in the commercial domain. They could be organizational boundaries, such as different branches of military service, different agencies, or different corporations. Or they could be system boundaries that were initially structured to align functionally or organizationally. In fact, the broader the scope of the mega-system, the more boundaries it will end up crossing. But crossing these boundaries also brings with it its own unintended consequence: multiple stakeholders and multiple owners, each of which has specific interests and equities that may align under certain circumstances but may, and do, conflict in others.

4. These mega-systems are rarely developed as a monolithic whole, but are *formed through* the process of *integration*; that is, they are "put together." Often, the components being integrated are in various stages in their individual life cycles and may have been developed using different standards and different design tenets.

5. The *constituent elements are, at least in part, independent systems* that have been developed to fulfill separately defined functions and continue to do so

even when detached from the whole. Of special importance: the further up a system is in the hierarchy of systems (see Figure 3.1), the more likely it is that the constituent elements are independent systems, independently developed.

6. These systems often have a *significant human, organizational social dimension* that contributes both to the complexity of behavior and to the evolution of the mega-system, both while its being developed and after it is put into operation.

3.2.2 Emergence of Mega-Systems

Mega-systems do not all emerge in the same way. Some are created after the fact from already-existing systems. Others are deliberately initiated as formal efforts intended to provide a boundary-spanning capability. Still others are assembled in response to an immediate and urgent need or opportunity.

It is worthwhile to note that mega-systems may emerge along one path and then transition to another. For example, mega-systems may originate as an effort to integrate "legacy" systems and then transition to a more formal integrated effort as sponsors, users, and developers encounter difficulties in integrating systems that were separately developed. Similarly, boundary-spanning solutions that are developed to meet an immediate need may later transition to formal acquisition efforts. In other cases, where the integration efforts proved successful, these may spawn other similar efforts to bridge separate systems.

3.2.2.1 "Composed" Mega-Systems

"Composed" mega-systems are those large-scale, complex systems that are formed from the integration of *previously* developed systems. This class of mega-systems most often comes to mind under the label "system-of-systems."

This class of mega-systems has two distinguishing characteristics. First, the component elements were initially developed separately to meet their local requirements, with no expectation that they would operate as part of a larger whole. Second, the mega-system is rarely formally structured as a single acquisition program. That does not imply that there were no requirements to interface (exchange information) with "external" systems. In fact, there is typically a long list of external interfaces specific to both the external system and the data/information exchange. What it does mean, however, is that there was no a priori concept—or vision—of the mega-system or of the outcomes that its operation could achieve. Instead, that concept or vision emerged only after the systems started (or in some cases completed) their separate developments. The resulting challenge is to integrate systems that were not individually designed to facilitate integration and, at the same time, to coordinate the efforts of separate program developers who have may have no particular incentive to collaborate.[3]

While the component elements are often formal programs of record with their own constituencies, requirements documents, developers, schedules, and funding, the composed mega-system typically lacks these. In the absence of formal authority, organizational mechanisms, or funding, the composed mega-system must often rely on influence rather than directives to define common approaches. Frequently it takes an overarching goal or a common threat before the separate components are able (or willing) to act in concert.

One example of a composed mega-system is the Single Integrated Air Picture (SIAP), a project initiated to solve some long-standing problems related to track data for theater air and missile defense applications (see Chapter 7). While the participating systems all used the same data link standards, their implementations were sufficiently different that the results could not be readily integrated to yield common and unambiguous tracks of all airborne objects in the surveillance area. To reach this goal, the DoD established the SIAP System Engineering Task Force (SETF) to fix recognized problems in the existing network and to guide development toward a future capability.

A second example of such a mega-system is Theater Battle Management Core Systems (TBMCS), a U.S. Air Force program that was initially structured to consolidate three legacy systems into a single, integrated command and control system (Figure 3.2). TBMCS was intended to provide a joint air operations center and its theater components with a common and shared air operations and intelligence

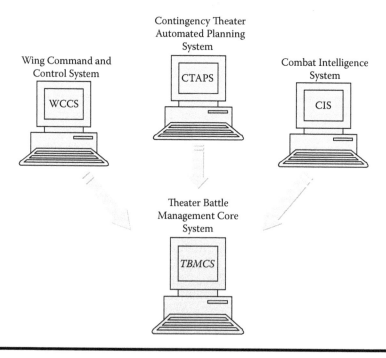

Figure 3.2 Integration of legacy systems into Theater Battle Management Core Systems (TBMCS).

database, as well as common software tools for planning, executing, and sustaining air war campaigns.

Rather than develop a new system, the initial task was to integrate the legacy systems using commercial-off-the-shelf (COTS) information technologies. The expectation was that the distinctions between the different components would disappear over time. It is important to note that although TBMCS was structured as a formal program of record, the Air Force did not establish an overarching concept of operations for the integrated system, nor did it produce a new requirements document. Instead, for a variety of reasons, the initial strategy was to use the existing requirements documents governing the legacy systems. Consequently, the requirements were expressed in terms of legacy system functionality. Despite encountering many challenges and failing its first operational test, TBMCS passed its second test event and received a favorable fielding decision. Since then, it has fielded several spirals of capability and is now deployed as the mandated system used to plan, manage, and execute the Air Battle Plan.[4]

3.2.2.2 "Designed" Mega-Systems

"Designed" mega-systems are also large, complex systems, but what differentiates them from other mega-systems is that they are established to develop a new capability from a more-or-less clean sheet. This strategy is often a reaction to both technical and management problems encountered in trying to integrate separately developed systems. Because of their scale and the number of separate systems involved, these designed "mega-systems" can be massive undertakings.

The designed mega-system can be managed as a single program or as a family of cooperating programs. As formal programs, designed mega-systems have a designated program manager with funding authority, formal requirements documentation, and a centralized systems engineering authority. Statutory and regulatory acquisition policies, such as independent operational testing, apply unless specifically waived.

What makes them mega-systems rather than just large monolithic systems is that the components are themselves systems, some of which already exist, others of which must be developed specifically for the particular effort. Because of the scale of the effort, the components are often acquired from multiple contractors, each specializing in a particular class of component. Consequently, much of the technical focus is on the interactions between these quasi-independent components, and integration is the major technical activity and a well-recognized risk to be managed.

An example of a "designed" mega-system is the U.S. Army's Future Combat System (FCS) program (Figure 3.3), which does significantly more than merely replace existing heavily armored ground combat systems with lighter, more readily transported vehicles. Instead, it is intended to be the cornerstone of the Army's strategy of transformation into what is termed the "Future Force." The FCS

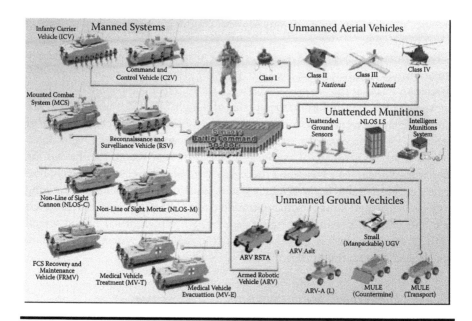

Figure 3.3 The U.S. Army's Future Combat Systems (FCS).

collectively consists of a family of advanced, networked air and ground systems that will include both manned and unmanned platforms. Key to the integration of these individual platforms is a suite of information technologies; networked reconnaissance, surveillance, and target acquisition capabilities; and battle command systems. Unlike previous efforts in which weapon platforms were separately developed and separately fielded, this effort is directed at developing an integrated capability for a Brigade Combat Team.

A second example of a designed mega-system is the Distributed Common Ground/Surface System (DCGS). This is a cooperative effort between the individual military Services, DoD agencies, and intelligence agencies to enhance real-time networking of Intelligence, Surveillance, and Reconnaissance (ISR) systems, improve access to ISR information by operational users in the field, and increase interoperability among ISR systems. Programmatically, DCGS is structured as an overarching family of interconnected systems for collecting, posting, processing, exploiting, and disseminating ISR information. It also serves as the DoD's "hub" to effectively implement the information sharing relationships between the warfighters, the intelligence analysts in the individual branches of the armed Services, and the various intelligence agencies.[5]

While each of the military Services is developing a particular variant (Figure 3.4), they are all committed to making the versions interoperable via a common information infrastructure known as the DCGS Integration Backbone (DIB). The U.S. Air Force is developing this common infrastructure for use by all the Services, and a DoD-level governance process has been established to ensure compliance and evolution.

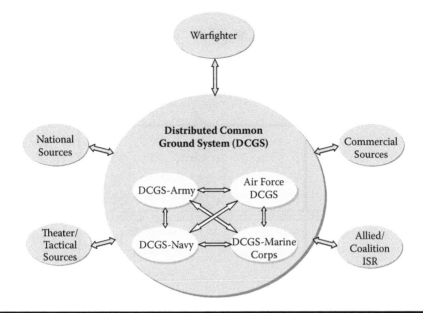

Figure 3.4   Department of Defense Distributed Common Ground System (DCGS).

3.2.2.3 *"Dynamically Assembled" Mega-Systems*

"Dynamically assembled" mega-systems are assembled and integrated to meet an immediate operational need or opportunity. They may consist of systems of record along with commercial or government off-the-shelf components and operational prototypes. In many cases, they may encompass existing systems that are used in unconventional and totally unanticipated ways. There is little or no development activity. What development may occur is typically in the nature of "glue-ware": software or gateways used to integrate systems not previously linked.

The customers are typically the combatant commanders, and their immediate and particular problems are the focus of the activity. Each problem, and the solution, may be unique to the situation, but the systems often demonstrate a possibility that had previously not been considered. Successful dynamically assembled mega-systems often transition to more traditional engineering and acquisition activities or spawn other such efforts.

One example is the "cursor-on-target" rapid prototype (Figure 3.5) developed in response to a 2002 challenge to the Air Force community by the then U.S. Air Force Chief of Staff, General John Jumper, to develop net-centric operational capabilities to realize his machine-to-machine vision where "the sum of all wisdom is a cursor over the target."[6] The capability was designed to answer an Operation Northern Watch[7] request for a system that could rapidly feed accurate target coordinates from a combat controller in the field directly into the cockpit of a strike

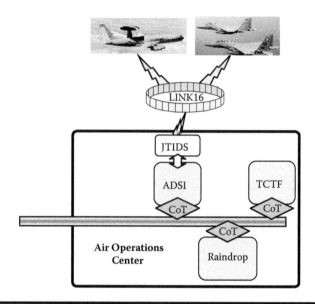

Figure 3.5   Initial U.S. Air Force cursor-on-target rapid prototype.

fighter aircraft (the F-15E). Until then, the procedure was to pass target by voice with all the attendant possibilities of error and delay.

The solution was a machine-to-machine approach (Miller, 2004) for moving a core set of target coordinate data (what, where, and when) from an intelligence workstation, called Raindrop,[8] via the existing communications link (Joint Tactical Information Distribution System, also referred to as JTIDS), to the targeting system aboard the aircraft).[9] This approach was extended to allow integration of other sources of targeting data, such as that generated by troops on the ground using a global positioning system (GPS) and a laser rangefinder.

A second example was the development of a capability to interlink existing command networks in support of operations in western Iraq. One network, Link 16, had already been installed on many of the airborne sensors and on F-15 strike aircraft. The other, the Situational Awareness Data Link (SADL), had been developed several years previously by the Air National Guard and Air Reserves as a means to display position location data of friendly ground troops in the cockpit of an aircraft providing close air support. Neither of these networks had been designed to interoperate with each other, nor had anyone conceived of a situation in which that would be required. Furthermore, SADL was originally designed to work with a specific ground force radio,[10] but the ground forces in western Iraq were primarily Special Operations Forces (SOF) who used a different suite of radios.

In response to this situation, the Air Force rapidly engineered a gateway to translate between these two networks. This gateway, known as the Battlefield Universal

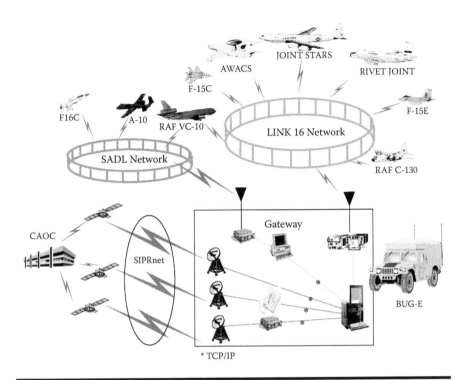

Figure 3.6 Gateway links separate networks.

Gateway Equipment (BUG-E), allowed data from ground forces, after approval from the SOF community, to pass over these two networks (see Figure 3.6). As a result, anyone on either network could share key pieces of information.

These two examples have a number of similarities. First, each focused on a specific problem that was important to the operational user. Second, each was able to provide operationally useful solutions in very short order, typically in a matter of months. The solutions did not entail substantial new development, but rather the integration of existing capabilities in innovative ways. Third, each required the active and purposeful collaboration of multiple organizations, including operational users as well as the engineering and acquisition elements, which were not within a single management structure.

3.3 Summary

Mega-systems are already prevalent in today's military, government, and business domains, and new ones are being defined and developed. Consider the next-generation National Airspace System being defined by the Federal Aviation

Administration (FAA) along with its partners in the Departments of Transportation, Commerce, Defense, and Homeland Security, the National Aeronautics and Space Administration, along with commercial and private interests. Consider also the technology portion of the Department of Homeland Security's Secure Border Initiation, referred to as SBInet. It is intended to integrate technology, staffing, and response platforms into a single, comprehensive border security suite. In the commercial world, consider the numerous business-to-business that enable transactions and information sharing between strategic partners. The development of the Electronic Product Code for use across the global supply chain, highlighted in Chapter 8, is an example of such a commercial mega-system.

Thus, systems engineers can expect that more and more of the systems they build today will be aggregated into ever-larger mega-systems. Furthermore, as the military, other government agencies, and commercial domains seek to leverage the power of shared information and knowledge, one can only expect that an increasing number of new, large-scale programs will be initiated to facilitate information sharing across previously separate organizations. Examples abound in the popular press and in trade journals. Federal, state, and local organizations are creating shared systems to counter the terrorism threat. Global corporations are creating e-business solutions and are using shared information systems to do so.

The next chapter presents approaches to aid in formulating a conceptual understanding of mega-systems. Thereafter, we provide some practical guidelines to help systems engineers deal with the special demands posed by these new types of systems.

Endnotes

1. In 2009, the FCS program was restructured, accelerating deployment of some components and terminating the manned ground vehicle portion of the program. Subsequently, the Army announced plans to develop a new ground combat vehicle concept, incorporating lessons learned in recent operations.
2. A significant literature on systems theory and complex adaptive systems can provide additional detail for the interested reader. Good sources are the Santa Fe Institute (http://www.santafe.edu) and the New England Complex Systems Institute (http://www.necsi.org).
3. For a summary of these concepts, see Faughn (2002).
4. For a detailed description of the systems engineering processes used to produce the first version of TBMCS, see Collens and Krause (2005).
5. Statement of Dr. Stephen A. Cambone, Under Secretary of Defense for Intelligence, before the Senate Armed Services Committee, Strategic Forces Subcommittee, 7 April 2004.
6. From Paone, C. 16 September, 2009. Hanscom to Host Cursor on Target Users Meeting Next Week. http://www.afmc.af.mil/news/story.asp?id=123168085 (accessed 6 December 2009).

7. Operation Northern Watch was charged with enforcing the no-fly zone north of the 36th parallel in Iraq and monitoring Iraqi compliance with United Nations Security Council resolutions.
8. Raindrop is a stereo imagery exploitation tool for precise coordinate extraction. It lets operators "view" target areas with three-dimensional imagery and produces weapon-planning target coordinates that can steer precision-guided munitions.
9. JTIDS is a network radio system used by the U.S. armed forces and their allies for data communications, principally in the air and missile defense community.
10. SADL was initially designed with the assumption that ground forces would be equipped with Enhanced Position Location Radio Systems (EPLRS).

Chapter 4

A Framework for Exploring Mega-Systems

The previous discussion and accompanying brief snapshots of mega-systems highlighted characteristics along two separate dimensions: technical and management/decision making. These vectors can also be thought of as representing two fundamental sets of processes. The first are the engineering processes required to understand, predict, and optimize the functions and behavior of the system(s) of interest. The second are the management processes involved in defining the goals and objectives to be pursued and the most efficient allocation of resources to achieve these goals and objectives. The management dimension focuses on the extent to which an agreed-upon view of the whole drives decisions about its parts. To further complicate matters, these two sets of processes apply in the broader environment in which the system is being developed and in which it will operate and, in some cases, continue to evolve.

Building on these dimensions, we propose a basic framework to help in

- Understanding the characteristics that distinguish mega-systems from other, more traditional notions of systems
- Exploring and extending our understanding of how to engineer and acquire them

4.1 Basis for the Framework

This framework builds on two fundamental tenets. The first recognizes that projects that require systems engineering, even those of similar types, exhibit fundamental

55

differences that can be identified and described. In fact, we believe that a project should be characterized along multiple dimensions, including not only what is being engineered, but also the context in which it is being engineered and acquired and the context in which the resulting system will operate or be used. The second tenet builds on the first. It asserts that these fundamental differences among systems and projects may warrant different engineering and acquisition strategies and approaches.

> Differences among systems and projects warrant different engineering and acquisition strategies and approaches.

At the core of these two concepts is the basic notion that the way we approach the engineering and acquisition of these projects has to take into account their underlying characteristics. This is not a new idea. In fact, we have frequently acknowledged that "one size does not fit all." This, we strongly believe, applies not only to the management of large projects, but also to the approaches used in engineering them.

Similar ideas have been offered in different fields. In the field of operations research, Jackson and Keys (1984) argue that system-based problem-solving methodologies should be selected based on the context of the problem at issue. To help in choosing the methodology, the authors offer a classification scheme that takes into account two key dimensions of the problem context: (1) the nature of the decision makers and (2) the nature of the system itself. In effect, the authors define a 2×2 matrix and go on to propose different operations research techniques that are best suited for each cell of the matrix.

In the field of project management, Loch and co-workers (2002, 2005, 2006) focus on the management of what they call "novel" projects. They characterize projects along two dimensions—project uncertainty and project complexity—and assert that the project management techniques that work well with predictable projects are ill-suited to those that exhibit high uncertainty and high complexity. Novel projects, they contend, are fundamentally different from risky projects and warrant different strategies, different project manager mind-sets and styles, and different management techniques and management infrastructure. They suggest that traditional planning-driven project management and risk management methods are most appropriate for projects of moderate complexity, where the nature of the "solution space" is known and that operate on "known terrain." For novel projects exhibiting high uncertainty and high complexity, these techniques do not suffice and, in some instances, will actually be counterproductive. For these types of projects, these authors propose two alternative project management approaches: (1) learning and (2) selectionism.

Loch and co-workers (2006) define learning as "the flexible adjustment of the project approach to the changing environment as it occurs; these adjustments are based on new information obtained during the project and on developing new—that is, not previously planned—solutions during the course of the project." Learning relies heavily on experimentation, external input, and close customer interaction. Selectionism, on the other hand, entails generating sufficient variety by running several alternative sub-projects to determine the one that yields the most desirable outcome. It entails considering several alternatives and confirming that solutions are feasible before committing to them.

In the field of engineering management, Shenhar and co-workers (1996, 1998, 2001, 2007) and Dvir and co-workers (1998, 2003) also advocate the use of a multidimensional framework for distinguishing among projects based on their levels of uncertainty, complexity, and pace. They relate uncertainty to the novelty of a product in the market or to economic or political environments. Internal uncertainty can arise from the introduction of new technologies or new features. Other sources of uncertainty include the levels and types of skills available, as well as the organizational culture. Complexity, from their perspective, is related to system scope; they rank projects on a range from efforts that develop system components (they refer to them as assembly projects) to system projects that produce an integrated result. Most complex, according to their taxonomy, are array projects, which are not unlike the system-of-systems described above. Pace refers to the time criticality of the project, ranging from regular projects through time-critical ones and, at the extreme, to "blitz" projects. Like the other authors cited above, they argue that one approach does not fit all instances, and that projects with different characteristics warrant different management and organizational styles.

While these authors focus on different disciplines, their recommendations have one major element in common. Rather than a single, well-defined approach that can be applied universally, each offers a rich view of a "toolkit" or "arsenal" of management techniques, practices, and even management styles that is available to the engineer and project manager. The obvious challenge then becomes that of understanding the particular circumstance of the project and its context, and to match the tools and techniques to the particular situation.

The notion of defining the problem context along multiple dimensions provides the conceptual basis for the framework. The concept of matching problem-solving techniques to the particular problem context underlies efforts to understand which processes and techniques of traditional systems engineering still apply to the world of large-scale, complex systems and to initiate the process of defining new ones where it becomes apparent that these are needed.

4.2 Elements of the Basic Framework

Figure 4.1 presents the elements of this framework as a set of dimensions or vectors that emanate from the origin at the lower left-hand corner. Three dimensions are shown. Along the horizontal or *x*-axis, we depict the dimension that characterizes the behavior of the system itself, ranging from simple and linear closer to the origin and becoming more complex as one moves further away. Along the vertical or

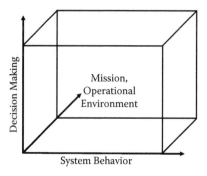

Figure 4.1 A simple framework for exploring mega-systems.

y-axis, we depict the management or decision-making vector, ranging from unitary to pluralistic. Along the third dimension, the *z*-axis, we depict the environment in which the system is expected to operate. That environment can range from stable and predictable near the origin to increasingly fluid and evolving. Thus, as one moves out from the origin in all three dimensions of the framework, one encounters increased *complexity*, *diversity*, and *uncertainty*.

4.2.1 The System Behavior Vector

Along the system behavior vector, systems are distinguished according to their degree of complexity, ranging from simple and linear to complex. As Jackson and Keys (1984) and others describe them, simple systems, otherwise termed "linear" or "mechanistic" systems, are characterized by regular, well understood, and, to a large extent, predictable behavior. In general, these are recognized as following well-established rules of behavior, such as the laws of physics or mechanics. Linear systems are relatively closed to the environment, in that events external to the systems do not significantly affect their behavior. Finally, their component elements are not purposeful; that is, the elements exist only as part of the larger system and do not follow their own independent goals. For example, an aircraft's engine is clearly a component element of the aircraft, but has no independent function when removed from the aircraft. By contrast, a weapon system that operates as part of a larger unit can still function independently when isolated, albeit likely with lower effectiveness.

Simple, linear systems are predictable, regardless of scale, because they exhibit the features of linearity; that is, they are proportional, additive, capable of being replicated, and they demonstrate causes and effects. By *proportional* we mean that small inputs result in small outputs, and proportionally larger inputs result in proportionally larger outputs. Understanding the relationship between input and output at one scale allows us to understand the same relationship at successively larger scales. By *additive* we mean that the whole is equal to the sum of its parts—no more and no less. By *capable of being replicated* we mean that the same stimulus under the same conditions will produce the same reaction, no matter how many times it is repeated. Finally, it is possible to *demonstrate cause and effect* by observation, calculation, or inference.

> Therefore, the nature of linear systems is that if you know a little about their behavior, you know a lot. You can extrapolate, change scales, and make projections with confidence. Unlike nonlinearity, in which 2 + 2 may yield oranges, in linearity you can rely on 4.[1]

Note that nothing has been said about the size of the system. Simple, linear systems can, in fact, be very intricate, have many component parts, and be quite large. Simply put, what distinguishes simple, linear systems from complex ones is not their size or even the number of components, but what we understand and can predict about their behavior.

Complex, nonlinear systems can also have a relatively large number of richly interconnected and well-interrelated elements. More important than mere scale, however, is the behavior they exhibit. First, not all the attributes and behaviors of the system are directly observable and, where observable, not all the interactions are understood. Second, they do not follow well-ordered, predictable rules of behavior. Third, complex systems exhibit emergent behavior, in that the interaction among components results in behavior that is unexpected and sometimes quite different from the behavior of the components themselves. Solutions to specific problems may well result in totally unexpected responses in different part of the systems or at different times. Thus, it may be difficult to predict the effects of a change without actually implementing it. Finally, complex systems can adapt; that is, they can interact with their environment and thus evolve over time.

Complex systems cannot be understood merely by decomposing them into their constituent elements and analyzing these elements separately. Such decomposition makes it possible to understand how the individual parts work, but not necessarily how they work together. Instead, understanding must focus on the interactions of the parts with each other, with the mega-system as a whole, and with the still larger system in which the mega-system participates.[2] This forms, in effect, a three-level perspective in which the interactions between the system and its "containing" system are as important, if not more so, than interaction among the parts of the system.

Along the continuum from the more simple and linear to the more complex, one can line up systems in the order in which human activity dominates (Figure 4.2).

Figure 4.2 System complexity continuum.

The most linear systems are the most machine-like. They follow well-understood laws of behavior based on physics and mechanics. While humans can and do interact with the machine-like systems, that interaction is typically kept to a minimum or otherwise routinized through extensive training to ensure that responses are consistent and hence behaviors are predictable. For example, airplanes are designed to minimize unexpected behaviors, and airplane pilots are trained to respond in specific ways to specific events.

Next along the continuum are information systems. These provide consistent, structured data or information about a particular area of interest. Human interaction with these systems is typically limited to queries and reports. Any search on the Internet will yield hundreds of examples of such information systems, including systems designed to track health statistics, foreign trade data, and geographical information. Such information systems are deliberately designed to limit the range of acceptable inputs and outputs, and will notify the user when necessary data has not been entered or has been entered in the wrong format.

Still further to the right in Figure 4.2 are cognitive systems. Unlike information systems, which are, in essence, look-up tables of data of interest, cognitive systems are intended to interact with users and support them in activities such as planning, analysis, and decision making. Examples of such systems are situation awareness displays, route planning aids, and other tools to assist in the analysis of alternative courses of action. Because these systems support problem solving, they must be able to address the range of problems that their users encounter and must also accommodate different cognitive styles and decision-making approaches. However, these types of systems, while more complex, still are limited in that the definition of the types of problems to be addressed is often designed up-front. Changes in the dimensions of the problem, or the introduction of new problems beyond those that had been initially considered, can render these systems less useful to the decision maker.

At the far right of the continuum are systems that deal with the interactions between social groups, whether these groups are units from different military Services or from different government agencies, corporations, or even nations. Examples are systems that facilitate collaboration among partners in the supply chain or among coalition members in a military operation. Jackson and Keys (1984) note that problems with these more complex systems inevitably involve "behavioral" issues because these systems are affected by political, cultural, organizational, and even ethical issues. They also note that changing values are important sources of change in these systems.

Thus, we assert that the systems with the greatest complexity as well as the greatest potential are those that involve interactions among different communities of interest with distinctively different modes of operation, languages, and other characteristics. Logically, then, these systems pose the greatest challenges to systems engineers.

Complexity, as we have discussed it, can manifest itself both in the development of the system and during its operation. During development, it may prove difficult to build a sufficient understanding of its full range of behavior. This may be because development is distributed over different subcontractors or occurs in different geographical locations. Modeling capabilities may be limited or otherwise insufficient to help developers understand all possible internal interactions. Or, particularly in the case of systems-of-systems, different components are developed asynchronously and without the expectation that they would be required to interact.

When dealing with mega-systems, systems engineers may work on solving problems in one area, only to recognize later that the proposed solution has had unintended consequences in another area of the system. They may find that in actual operation the system is used in fundamentally different ways than was initially anticipated and that, in practice, it may not function as it did in the controlled environment of the laboratory. This may occur as soon as the system is launched as a result of unanticipated initial conditions that generate wide-reaching behaviors, or it may not be evident until the system scales up from its initial limited implementation. Complexity may originate within the system and result from the unanticipated interactions of its parts, or it may result from the system's interactions with systems in its larger environment. What is predictable locally may not, in fact, be predictable globally. Thus, one of the challenges of engineering mega-systems is that while many of the uncertainties inherent in their design and operation may be foreseen, others may be unforeseen. These are the unknown unknowns.

4.2.2 The Decision-Making Vector

The management or decision-making vector is plotted along the vertical axis in Figure 4.1. Here, the continuum ranges from unitary decision making near the origin to pluralistic decision making at the other end of the axis.

Unitary decision making implies that all the stakeholders agree as to the goals and objectives of the system. Consequently, individual stakeholders make and implement their local decisions consistent with and in accordance with these common goals. Unitary decision making can be voluntary or can result from centralized direction, presuming, of course, that individual decision makers actually follow such direction.

In contrast, the decision-making context is pluralistic if it is characterized by competition about the goals and objectives to the achieved, the approaches to be followed, the particular features to be implemented, and the resources to be expended. In such cases, stakeholders can experience recurring periods of open, or perhaps more subtle, conflict. Decision makers will be apt to focus more on their local interests and concerns and be less interested in and less willing to support the achievement of more global objectives. In such instances, the few decisions that can be made will address only those aspects on which the various stakeholders can, in fact, reach agreement. On occasion, decisions can be imposed on the stakeholders by a higher authority, but in these cases they can be subject to pushback, ranging from overt and blatant noncompliance to more covert and subtle resistance. "Creative noncompliance" exemplifies just such a subtle response. Here, decision makers do not necessarily refuse directly to implement decisions that they feel may have been imposed on them and with which they do not agree. Instead, they may give the appearance of concurrence and compliance but may actually do the minimum amount that they feel they can "get away with." In either scenario, progress toward global solutions is, at best, slow.

4.2.3 The Context or Environment Vector

The third dimension is that of the context or environment in which the system will operate. Closer to the origin, one finds environments that are stable and predictable. In these environments, the tasks to be accomplished, the interactions among participants, and the information flows are well understood and, more importantly, are not likely to change much over the time it takes to develop and field the system. We refer to these as stable environments. As an example, consider the task of generating paychecks. This is likely to remain largely unchanged, despite the prevalent shift from paper checks to direct electronic deposits; consequently, design decisions made on the basis of today's business processes will probably continue to apply for some time to come.

At the other extreme, the context or environment can be highly fluid. In these cases, the people, processes, and information flows are subject to considerable, possibly volatile changes. New business processes can arise and unexpected partnerships can be forged, some of them enduring and others more fleeting in nature. As examples, consider the relationships among business partners in e-business or between organizations at various levels of government that are charged with counterterrorism. In such circumstances, it is difficult to anticipate the characteristics of the specific environment on the basis of an understanding of today's ways of doing business. Consequently, basing engineering decisions on an understanding of today's business process may prove problematic.

4.2.4 Region of Well-Bounded Systems

Systems whose behavior is linear and predictable, that have agreed-upon goals and objectives and a well-understood and stable mission space, are termed "well-bounded systems" and occupy the lower-left region of the framework (Figure 4.3). This can also be considered the domain of the traditionally ideal project. After all, what program manager would not prefer to head a project in which the requirements are known and not subject to change, the funding matches what is to be built and remains stable throughout the life of the project, and the technologies are mature?

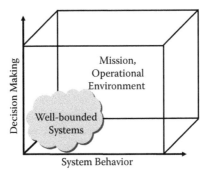

Figure 4.3   Region of well-bounded systems.

Well-bounded systems, therefore, are those that have:

■ Well-defined boundaries that differentiate the system of interest from the larger, "containing" system or environment
■ A reasonably stable, persistent operational environment
■ A set of agreed-to requirements that can be well defined, are precisely stated, and are expected to be stable over time
■ A set of functions that can be decomposed and allocated to the component elements with the expectation that when they are subsequently integrated, the overall behavior of the system will be as expected
■ A unified management structure

It is these well-bounded systems that best lend themselves to traditional systems engineering and development approaches. Checkland (1978) has termed these approaches "hard systems thinking." They include classical operations research, systems engineering, and systems analysis, and are based on "the assumption that the problem task they tackle is to select an efficient means of achieving a known and defined end." Because of the linear nature of the system's behavior, the engineer can more readily predict the technical interactions of the system's component elements and therefore has greater control over them. Moreover, because there is at least written agreement as to goals and objectives, the manager can make decisions to maximize the achievement of these desired outcomes.

It is worth pointing out that managers of traditional programs spend considerable energy trying to shape their programs to make them into such well-bounded systems. They define the boundaries of the program to encompass those elements over which they do have control and exclude or defer those elements over which they lack control. They structure the interfaces across these boundaries and formally manage them. They seek to minimize their dependence on components over which they have little control, to contain external influences over their system and to minimize perturbations to their requirements baseline. Requirements "creep"— a term that refers to changes to the requirements baseline—must be avoided; or where that is not possible, must be minimized and controlled.

4.2.5 Region of Mega-Systems

Mega-systems, in contrast, fall to the right of and above these well-bounded systems (see Figure 4.4). In some cases, they also encompass them: that is to say that some aspects of mega-systems are, in fact, well-bounded.

Mega-systems, as we are beginning to understand them, are characterized by:

- Requirements that are often stated as vision statements or broad architectures. These requirements evolve in response to changes in the environment, in user expectations, and in the technology base.
- Some functionality that emerges from the interaction of the components themselves without specific direction. That is, it is neither engineered in nor engineered out.
- The need to manage uncertainty—both downside risks and unanticipated opportunities.
- The need to deal with competition not only for resources, but also for alternative solutions, because the systems often cross program boundaries.

Mega-systems can exhibit complex, nonlinear behavior for all the reasons discussed above. Because they are often first-off systems, they lack the predictability gained from past experience. That knowledge may exist for parts of the system but rarely encompasses its totality. Further, they often rely on technologies that are still under development and whose performance can be estimated but not predicted with confidence. Similarly, the interactions between and among their components

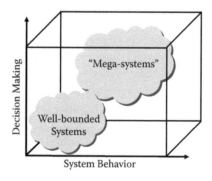

Figure 4.4   Region of mega-systems.

and between the systems and the user community(ies) is often based on a set of expectations that are, in effect, assumptions that have yet to be proven.

Mega-systems can also range from unitary to pluralistic. Sometimes they will emerge from agreement as to global goals, objectives, and actions. In such cases, stakeholders will make local decisions and take actions that further those goals and objectives. In other cases, such agreement can be perfunctory at best, or even non-existent. Alternatively, there can be creative noncompliance as stakeholders push back against unwelcome policies and directives. In such cases, stakeholders take actions to further their local objectives with little deliberate attention to the whole.

It is important to point out that not all mega-systems that have a single designated program manager are necessarily unitary. While there may be agreement on formal goals and objectives, stakeholders may—and do—have different perspectives, priorities, and constraints. This situation often occurs when stakeholders from different communities who have little common background or who have been competitors are brought together.

Similarly, one can envision situations where there is neither a single program focus nor a documented and approved set of requirements, and yet there is agreement as to stated goals and objectives. The Internet is such an example. It is a large, complex, loosely organized, collaborative mega-system consisting of autonomous, interconnected networks. It is maintained and evolves through the collaborative efforts of a global community. A large, open community of vendors and researchers develops and tests Internet standards. Compliance with these standards is voluntary rather than mandated; however, it is obvious that the value to be garnered from participation in the Internet far outweighs the costs of compliance with its standards. As each individual makes a local decision to comply with these standards, that local decision produces not only local outcomes, but also global ones. The greater the growth of the Internet, the greater becomes the value of participating in it.

4.3 Tame versus Wicked Problems

To a large extent, the distinction we make here between well-bounded systems and mega-systems maps well to the ideas set forth by in a seminal paper by Ritter and Webber (1973). Writing in the field of public policy, Ritter and Webber observed that there is a class of problems that cannot be resolved by applying traditional structured linear approaches. They call these "wicked problems" and distinguish them from "tame problems," which do lend themselves to solution by such traditional methods. Examples of tame problems include scientific problems and some types of engineering programs.

Multiple examples of wicked problems are cited in the literature and include

- Fighting terrorism
- Determining where to locate the highway

- Reengineering business processes
- Most public policy issues, including but not limited to national healthcare, immigration, and climate
- Developing a national identify card
- Designing and integrating complex software

While they can be quite complicated, tame problems lend themselves to reduction (breaking the problem into parts and solving each part individually). The process of developing solutions to tame problems can be organized into distinctive phases starting with problem formulation and proceeding to analysis and solution. It is possible to come to a solution in a reasonable amount of time. When this process has been completed, the problem solver knows when a solution has been reached and can judge its effectiveness. In the software engineering discipline, the Waterfall method is a classic example of a linear approach (see discussion in Chapter 5, Section 5.1.2).[3]

Wicked problems, in contrast, do not lend themselves to resolution by such methods (Conklin, 2006). Their boundaries are more difficult to delineate, and their root causes are more difficult to identify. They are characterized by complex internal and external interdependencies. Wicked problems can perhaps be best characterized as those problems in which stakeholders do not agree, and both requirements and constraints are likely to change.

> Wicked problems are ones in which stakeholders do not agree, and both requirements and constraints constantly change.

In dealing with wicked problems, problem formulation and the solution formulation are intertwined. The process of defining the problem helps to define the solution options; and similarly, the process of articulating potential solutions helps frame the problem. Solutions to certain aspects of a wicked problem can uncover or even generate complex problems elsewhere. And, just as one cannot understand a wicked problem independent of its context, solutions to wicked problems can generate unpredictable waves of consequences over an extended period of time.

When encountering a wicked problem, two approaches are possible. One can turn it into a tame problem and tackle it in the traditional manner. But, if that is not possible, then the worst approaches that one can use, according to Ritter and Webber, are to treat a wicked problem as though it were a tame one, tame a wicked problem prematurely, or refuse to recognize its nature as a wicked problem. Instead, a fundamentally different approach to problem solving is needed: one that is iterative, emergent, and collaborative, and is directed as much toward bringing the stakeholders to reach a common understanding of the nature of the problem and its possible solutions as it is to developing those solutions.

The notion of differentiating wicked from tame problems has been applied in various disciplines, most notably economic, environmental, and policy issues, and also in software development. DeGrace and Stahl (1990) point out that many of the systems problems facing software developers have all the characteristics of wicked problems. They view the Waterfall method of software development as an example of the type of linear methodology best suited to tame problems, but not at all well suited to the kinds of wicked problems that many software developments encounter. It is not so much the model itself with which they take issue, but rather the dogmatism with which it is applied.

With this context, we now move to a discussion of how the characteristics of mega-systems shape the systems engineering process. Because mega-systems often comprise linear and/or well-bounded systems, traditional approaches will continue to apply, but the layers of complexity created by aggregating them into mega-systems will require new approaches not only for technical execution, but also for technical management. The trick, of course, will be to recognize the difference between those parts of the problem that are "tame," and hence best suited to the traditional structured practices of systems analysis and systems engineering, and those parts that are inherently "wicked" and that demand new approaches.

Endnotes

1. See Czerwinski (1998) for a discussion of linearity and nonlinearity in military affairs.
2. See Ackoff (1993) for a discussion of the difference between analysis and synthesis.
3. The Waterfall Model is a top-down, structured methodology that consists of an ordered set of phases in which one phase serves as input to the next one. The phases are initiation, requirements analysis, preliminary design, detail design, coding, module test (unit test), system test, installation/delivery, and maintenance/enhancement. It is important to note that the classic Waterfall Model does not include feedback.

Chapter 5

Engineering and Acquiring Mega-Systems

In previous chapters we described mega-systems and highlighted some of their salient characteristics. We also proposed a simple framework that differentiates them from traditional, well-bounded systems. For several reasons, including their sheer scale, the nature and pace of change of their underlying technologies, the potential complexity of their interactions, and—perhaps most importantly—the fact that it is rarely a single organization that owns and therefore completely controls the mega-system, engineering these mega-systems entails new challenges. This chapter discusses the difficulties and drawbacks of applying traditional systems engineering processes and practices to the engineering of these massively interconnected, information technology-intensive mega-systems.

5.1 What Is Systems Engineering?

Systems engineering, as a discipline, emerged from the World War II experience of developing unprecedented weapon systems and was expanded in the post-war years to apply to large-scale technological problems in both commercial telecommunications and defense aerospace. Today, "systems engineering" has many definitions. For example, Martin (1996) defines it as

> ...the process that controls the technical system development effort with the goal of achieving an optimum balance of all systems elements. It is a process that transforms a customer's needs into clearly defined system parameters and allocates these parameters to the various development

disciplines needed to realize the system products and processes. Then, using the analytical SE methodology, the process attempts to optimize the effectiveness and affordability of the system.

According to Checkland (1978), systems engineering is, in essence, "the total task of conceiving, designing, evaluating, and implementing a system to meet some defined need – the carrying out, in other words, of an engineering project." (See also Checkland, 1981, 1989.)

Drawing on his experience at Bell Telephone, A.D. Hall (1962) documented the systems engineering methodology as a top-down problem-solving sequence in which a project team works toward a "single defined objective" (see Table 5.1). He went on to state that

> Systems engineering considers the content of the reservoir of new knowledge, then plans and participates in the action of projects and whole programs of projects leading to applications. Thus systems engineering operates in the space between research and business, and assumes the attitudes of both. For those projects which it finds most worthwhile for development, it formulates the operational, performance and economic objectives, and the broad technical plan to be followed.

In other words, systems engineering starts with a determination of *what* is to be developed and then proceeds to determine, in increasingly greater detail, *how* it is to be developed.

Table 5.1   Systems Engineering Process as Described by Hall[a]

Problem definition	Defines the need to be satisfied
Selecting objectives	Guides the search for alternatives; provides the criteria for selecting the optimum system
Systems synthesis	Identifies or creates alternative systems for consideration
Systems analysis	Compares alternatives with respect to objectives
Selecting the optimum system	Selects alternative that shows the most promise
System development	Design, construction, test, and evaluation of engineering prototypes leading to design freezing
Current engineering	Engineering activities during system use, including monitoring operations, extending the system to meet new objectives, and adapting the system to changing conditions

[a] Information taken from Hall, A.D. 1962. *A Methodology for Systems Engineering.* Princeton, NJ: D. Van Nostrand.

The International Council on Systems Engineering (INCOSE; INCOSE, 2004) posts the following definition of systems engineering on its website:

> ...an interdisciplinary approach and means to enable the realization of successful systems. It focuses on defining customer needs and required functionality early in the development cycle, documenting requirements, then proceeding with design synthesis and system validation while considering the complete problem:
>
> Operations
> Performance
> Test
> Manufacturing
> Cost & Schedule
> Training & Support
> Disposal
>
> Systems Engineering integrates all the disciplines and specialty groups into a team effort forming a structured development process that proceeds from concept to production to operation. Systems Engineering considers both the business and the technical needs of all customers with the goal of providing a quality product that meets the user needs.[1]

Hall's ideas and the more current INCOSE definition of systems engineering have several features in common. In both cases, they assume that a project is established to accomplish the effort. Related to that is the notion of a project team with clearly defined lines of authority. Third, both assume that precise objectives against which the system performance can be developed and assessed can be (and are) defined up-front.

5.1.1 Systems Engineering Process

Sage (2005) points out that

> Systems engineering is a process that is comprised of a number of activities that will assist in the definition of the requirements for a system, transform this set of requirements into a system through development efforts, and provide for deployment of the system in an operational environment.

Today, the systems engineering process (ANSI-EIA, 1999; ISO/ISC, 2002) is most often described using the "V" technical development model (Fosberg et al., 2000; INCOSE, 2004) and is based on a model developed initially by NASA. While there are many variations of the basic model, they all have the same essential elements (Figure 5.1).

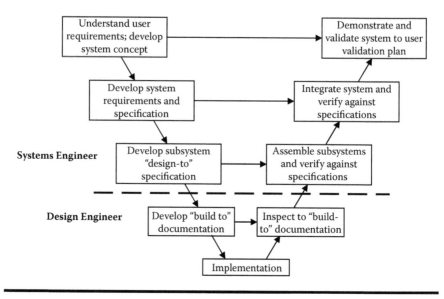

Figure 5.1 The "V" Model of systems engineering.

Time increases as you move from left to right in the model. In this instance, time includes not only project time, but also the sequence of events and the increasing maturity of the project. The left leg of the "V" depicts the process of decomposition and definition of the system to be developed. It starts, at the top, from the perspective of the user and then proceeds, in successively greater levels of detail, to specification of the subsystems that will make up the overall system. This process of decomposition allows components to be built by different subcontractors, all of whom, however, are obligated to build to the specification. The right-hand leg of the model represents the process of integrating these various subsystems into the larger system and verifying, at each step, that the results comply with the specifications. Just as decomposition and specification are the responsibility of the systems engineer, so too are integration and verification. The "V" Model thus describes a set of formal, top-down processes that depends heavily on documentation and a structured series of technical reviews and audits.

The types of projects in which systems engineering has proven most successful are typically those large efforts that provide a substantively new technical capability, using new technology that reflects the basic underlying principles and laws that govern the systems and for which goals and objectives are well understood and whose requirements are well defined and specifications are highly detailed (Bar-Yam, 2003). These are generally very complicated but essentially linear systems whose requirements are relatively stable, whose interfaces can be specified, and whose behavior is, to a great extent, predictable. In Ritter and Webber's (1973) lexicon, these are tame problems.

In his book entitled *Rescuing Prometheus: Four Monumental Projects that Changed the Modern World,* Hughes (1998) explores four illustrative systems engineering case studies: (1) the Semi-Automatic Ground Environment (SAGE) air defense network, (2) the Atlas missile program, (3) the ARPANET, and (4) the Boston Central Artery/Tunnel (CA/T) project. He contends that a new management style, based on the "systems approach," characterized the creation of these technological systems. While these systems were clearly technically challenging in terms of both their scale and the degree to which they were pushing the limits of the technology at that time, they—at least the defense systems—were "well focused projects with clearly delineated lines of authority." In contrast, significant and messy political, social, and environmental issues characterized the CA/T project, and the consequences are distressingly evident.

5.1.2 Software Engineering Processes

Software engineering[2] had its origins in systems engineering, and the two disciplines continue to be closely aligned. When a system consists of hardware, software, and people, software engineering is an element of the overall systems engineering process. In the kinds of mega-systems we describe, software is often the critical component.

Unlike systems engineering, which has one basic process model—although admittedly a model with many variations—the field of software engineering has produced a number of fundamentally different models. The most basic one is the so-called "Waterfall Model," which was originally described, although not yet so named, in a paper by Royce (1970). It became commonly known as the Waterfall Model because it is typically depicted as a series of steps in which the output of one serves as the input to the next (Figure 5.2).

Today there are many variations of the Waterfall Model, with different numbers of steps and different labels for each step. For example, the NASA Model assumes that software development is part of a larger system development activity, while the model for a stand-alone business application development may have to account for an additional project initiation step. Notwithstanding these variations, the Waterfall Model, like the "V" engineering model, is a formal model, predicated on the ability to

- Separate the "what" from the "how"
- Capture and document the complete set of requirements up-front with a reasonable expectation that they will not change substantially during the course of the project
- Defer integration and testing until the end of the process with the expectation that the pieces will fit together as designed and function as expected

While the Waterfall Model applies in some situations, it has a number of acknowledged shortcomings and does not work for all software developments (DeGrace and Stahl, 1990; Cantor, 2002). Key limitations include the following:

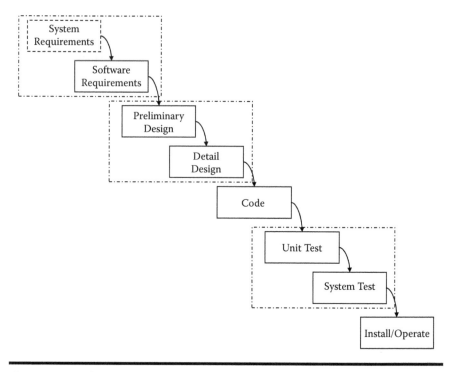

Figure 5.2   Software engineering Waterfall Model.

- It takes a long time and is expensive to use.
- By separating the "what" from the "how," it sets up a communications gap between the user and the developer.
- It assumes that requirements are stable and can be "frozen" for the duration of the development activity.
- It relies heavily on documentation and assumes that the documentation can be both complete and accurate.
- It works best for linear "tame" problems, but poorly for unstructured "wicked" problems that necessitate a more iterative or adaptive approach.

In response, the field of software engineering has created a number of alternative models, including incremental development, prototyping, spiral development, and adaptive development, that have evolved to address some of these well-recognized limitations of the Waterfall Model. Considerable literature is available on all these models and their variations. Therefore, it is not the intent here to describe them in sufficient detail to satisfy the software engineering professional, but merely to recognize that different models do exist and to explore whether these alternative models that were developed to address specific problems with sequential software engineering may suggest analogies for the systems engineering discipline.

Iterative and incremental development (Larmon and Basili, 2003) entails breaking the development into increments, or separate mini-projects, each one building on the capability provided in earlier increments. User experience with each increment provides feedback for subsequent enhancements and revisions.

Prototyping entails development of a working model to allow developers to demonstrate and test out parts of the design and to gain early feedback from users. In some cases, prototyping is followed by a more formal design process, while in others the prototype, if sufficiently mature, becomes the final product.

Spiral development is essentially an iterative development model that allows for interaction between user and developer within each increment. As described by Boehm (1988), each round progresses through the same steps (determine objectives, alternatives, and constraints; evaluate alternatives; identify and resolve risks; develop, verify next-level product; plan next phase). Each succeeding round calls for greater specificity, with the final round being what is, in effect, the Waterfall Model. This spiral model explicitly addresses risk and includes prototyping at each round as an approach to resolve risk.

Adaptive development (Highsmith, 2000, 2004) approaches represent a fundamental departure from the "big bang" approach exemplified by the Waterfall Model in that they reject the notion of preplanned development in favor of the ability to respond rapidly to changing circumstances. Adaptive development processes, such as Scrum, focus on short, time-bounded iterative developments in which working functionality is delivered at the end of each iteration, continuous collaboration between user and developer, and a marked tolerance for change. Scrum is one of several variants of adaptive software development and focuses on 15- to 30-day cycles (called sprints) and a short daily meeting (scrum) to review daily progress, plans, and impediments.[3]

Adaptive development has several key characteristics:

AGILE MANIFESTO

- Individuals and interactions over processes and tools
- Working software over comprehensive documentation
- Customer collaboration over contract negotiation
- Responding to change over following a plan

- Customer satisfaction achieved by rapid, continuous delivery of useful software
- Working software delivered frequently (in weeks rather than months or years)
- Working software, not documentation, as the principal measure of progress
- Regular adaptation to changing circumstances; even late changes in requirements welcomed
- Close daily cooperation between users and developers; communication ideally via face-to-face conversation
- Self-organizing teams

5.1.3 Emergence of System-of-Systems Engineering

Valerdi et al. (2007) present evidence of a debate occurring within the systems engineering discipline. Some experts believe that system-of-systems engineering is sufficiently different from the field of systems engineering that it warrants different processes, methodologies, and even tools (Keating, 2003), while others contend that traditional systems engineering practices do and will continue to suffice.

Certainly there is ample evidence that system-of-systems engineering, as distinct from traditional systems engineering, has attracted considerable attention. It has spawned new conferences (such as the *IEEE Conference on System of Systems Engineering*) and journals (such as the *International Journal on System-of-Systems Engineering*), as well as academic emphasis (such as the System-of-Systems Signature Area at Purdue University and the National Center for System of Systems Engineering at Old Dominion University). In the Department of Defense, system-of-systems engineering has been included in the *Defense Acquisition Guidebook* (DAU, 2006) and is the focus of an evolving *System of Systems Engineering Guide*[4] authored by the Office of the Under Secretary of Defense, Acquisition, Technology and Logistics.

Table 5.2, adapted from Valerdi et al. (2007), highlights some of the salient differences between systems engineering in a single project or program environment and systems engineering in a system-of-systems (or mega-system) environment.

5.2 Mega-System Challenges for Systems Engineers

For more than a decade, the DoD, in particular, has gained experience with efforts to engineer large-scale mega-systems, often through the integration of separate systems and, more recently, through the acquisition of what we have termed "designed" mega-systems. In doing so, the DoD has encountered several challenges, a few of which we briefly discuss below.

5.2.1 Increasing Scope and Complexity

What started out as a focus on designing an individual combat platform, such as a main battle tank or fighter aircraft, has evolved into the engineering of a complex mix of interacting ground systems or a constellation of aircraft. What started out as the engineering of a specific software application has evolved into the simultaneous engineering of all the applications that may be required for an operations center housing up to hundreds of staff. What started out as a closed mix of systems to accomplish a specific mission has been extended to a much broader capability that applies to an entire operational unit across a wide range of missions and operational environments. Mega-systems that embrace the entire enterprise are now common objectives not only for the DoD, but also for corporations and for entire departments or agencies of government.

Table 5.2   Systems Engineering and System-of-Systems Engineering

Systems Engineering	System-of-Systems Engineering
Focuses on development of a well-bounded system to meet documented needs and performance	Evolves a capability by integrating legacy systems and new developments; rarely a fresh start
System architecture established early in life cycle; expected to remain relatively stable throughout development process	Architecture dynamically adapts with changing needs and mix of component systems
Interoperability is formally defined and managed	Interoperability continues to evolve; some aspects still to be discovered
Centralized acquisition and management	Component systems (often) acquired and managed separately
Stakeholders associated with the program (developers, funders, users, sponsors, etc.)	Stakeholders associated with multiple programs; may entail competing spheres of influence
Established and widely taught discipline; well-defined and documented processes	Emerging discipline; processes still under development

Source: Adapted from Valerdi, R., A.M. Ross, and D.H. Rhodes. 2007. A Framework for Evolving System of Systems. *Crosstalk.* October 2007. www.stsc.hill. af.mil/Crosstalk/2007/10/0710ValerdiRossRhodes.html.

As the scope increases, the system grows to encompass not only greater numbers of similar components, but also a much richer mix of component types. The Army FCS, for example, includes a mix of different types of weapon platforms, ground and air robotic platforms, and sensors of various types, as well as the command, control, and communications that link the individual platforms into a network. This translates into more technologies, many of them critical to accomplishing desired goals, a broader mix of applicable engineering disciplines and specialties that must work together, and a larger and more varied set of stakeholders.

As the number of components grows, so does the possible range of system interactions among them as well as with the external environment. The very unpredictability of some of these interactions makes it difficult to engineer them solely within traditional specialty groups. Instead, effective engineering requires working from the perspective of the whole mega-system; but as the mega-system grows, it becomes increasingly difficult for any one individual or even small group to comprehend it in its entirety.

Still another dimension of scope and complexity deals with the two-way feedback between the mega-system and the "business rules" of the organizations that will use it. As the mega-system evolves, it necessarily changes how people use it and how they accomplish their essential missions. For example, the FCS will change how soldiers fight. At the same time, new ways of fighting will change what both operators and strategists expect from the system. Thus, the notion of having a well-defined, unchanging set of requirements against which to optimize the mega-system becomes questionable.

5.2.2 More Diverse Stakeholders

Mega-systems that encompass multiple functions or cross multiple organizations necessarily have a larger and more complex set of organizations and individuals who have an interest or stake in the development. The broader the set of stakeholders, the greater the likelihood that their interests will differ, rather than being congruent.

Consider the following situation confronting the DHS. Clearly, integrating 22 separate agencies and bureaus into a single department presents considerable challenges. Many of these organizations ran their own programs that now must be integrated and, where appropriate, consolidated. One such example is in the area of personnel screening and credentialing.[5] In 2005, the DHS proposed the creation of the Office of Screening Coordination and Operations, formally implemented in 2006 and renamed the Screening Coordination Office (SCO), to bring together eight separate screening programs (DHS, 2005). These programs, originally established by different agencies to meet specific needs and now operated by different elements of the DHS, include the following:

- United States Visitor and Immigrant Status Indicator Technology (US-VISIT)
- Secure Flight and Crew Vetting
- Free and Secure Trade (FAST)
- NEXUS/Secure Electronic Network for Travelers Rapid Inspection (SENTRI)
- Transportation Worker Identification Credential (TWIC)
- Registered Traveler
- Hazardous Materials Driver Background Checks
- Alien Flight Student Program

It is noteworthy that these programs are only a small part of the 28 DHS credentialing programs that were cited in a recent report. A report issued by the SCO in December 2006 lists 28 specific programs. In addition to the ones cited above, there are a number of programs used specifically for granting immigration status along with other initiatives such as the Western Hemisphere Travel Initiative[6] and REAL ID.[7]

By coordinating these separate programs, each with its own processes, constituencies, and systems, the DHS seeks to gain efficiencies. Meanwhile, business interests as well as individuals, including individuals seeking to travel to the United States, hope to simplify the process so that they do not have to work with multiple bureaucracies, each with its own system and procedures or have to resubmit their information and pay multiple times for the same background checks.

At the same time, privacy right groups are concerned that such consolidation will increase the threat to individual rights. Here we see a case where expanding the scope of the effort may benefit some stakeholders—in this case the Department, some travelers, and business interests—but may raise suspicions among other interests.

The situation is considerably more complex than even the challenge of coordinating, let alone consolidating, multiple systems in the same department. Other interests also have a stake in the screening and credentialing "enterprise." While the DHS is concerned about securing borders, protecting transportation systems, facilitating legal immigration and trade, enforcing immigration and trade laws, granting immigration status, and protecting the infrastructure, other federal agencies, state and local governments, economic interests, and advocates also have a stake in this area. For example, the U.S. State Department grants visas, the Bureau of Prisons and the U.S. Marshal's Service transport prisoners, and the Intelligence Community gathers and analyzes intelligence.

State and local governments are concerned with law enforcement,[8] administration of welfare programs, and other services to their constituents. They are also concerned about having to pay for unfunded mandates, such as REAL ID, that are placed on the states by Congress but are perceived as DHS requirements. Economic interest groups, such as unions, shippers, and employers, are concerned with maintaining wages and jobs, and seek to balance those interests with concerns about efficient traffic movement and compliance with labor laws. Finally, advocacy groups concerned with immigration and privacy issues, among others, also have a stake.

This synopsis highlights the complex scope of screening and credentialing as a cross-enterprise challenge. It is being done in the context of the largest government agency consolidation and reorganization since the formation of the DoD in 1947. The technical aspects of coordinating multiple systems are not trivial by any means, but they seem to dwindle in comparison to the political, organizational, operational, economic, and even cultural challenges of accommodating such a wide range of diverse and—in some circumstances—competing internal and external stakeholders. Further, this effort must be accomplished while the different agencies continue to exercise their missions.

Engineering systems of this scale, whether the systems are developed under a single program or through coordination and consolidation of multiple programs, requires attention not only to the most effective *technical* solution, but also to these other, less technical but no less vital, dimensions. As George Heilmeier, Chairman Emeritus of Telcordia Technologies, has said: "The hard stuff is the soft stuff."

5.2.3 Multiple, Often Conflicting, Objectives and Constraints

Few existing institutional incentives encourage program managers to build cross-program capabilities at the expense of or in contravention of the capabilities required to satisfy their own constituencies. Consequently, the components may have individual goals and constraints that differ not only from those of other components, but also from the goals and constraints of the larger whole. This obviously can occur when separately developed systems are integrated during or after their development. It can also occur even within the bounds of a single program effort when the components are individually associated with discrete communities of interest that have fundamentally different perspectives, interests, and inherent "cultures."

For example, one large-scale simulation system was intended to meet the training needs both of the joint force and the individual armed Services. However, the various participants were unable to reach agreement on the capabilities to be provided. The tension between joint and individual service requirements adversely affected the expected product capability and the development schedule. The military Services expected more detailed modeling of their operations than joint training requires. Thus, advocates of joint training saw the service simulations as needlessly complex and expensive. On the other hand, the advocates of service training found little utility in simulations built only to joint requirements.

After two reorganizations, the DoD leadership elected to try a continuous integration strategy. The challenge of continuous system integration is to design a plan that yields a sequence of growing, testable capabilities and thus avoid the "big bang" problem, where all system components must be integrated before any capability can be tested. "Big bang" integrations find system problems after the development phase is complete, when the resources needed to fix the problems are harder to obtain. The continuous integration approach proved very effective for the overall system, but brought an increased reliance on precise schedules containing a large number of incremental, component deliverables. Again, the tension between joint and Service requirements complicated development. Individual military Services held differing views about the order in which their capabilities were to be developed, and were often unwilling to tolerate adjustments of their internal schedules for the greater good of the overall system integration schedule.

5.2.4 Dissimilar Time Scales

When people refer to the rapid pace of technology change, they are typically referring to information technologies. As noted in Chapter 2, we are familiar with several "laws" of technology, most notably

- Moore's law, which states that the power of computers at the same price doubles every 18 months
- Law of fiber, which states that bandwidth capacity at the same price doubles every 9 to 12 months
- Metcalf's law, which states that the power of the network goes up with the square of the number of people connected

But not all technologies change at the same pace. Material and sensor technologies, such as advanced materials and ceramics, autonomous robotics, and chemical and biological detection capabilities, evolve at considerably slower rates than information technologies, typically over a period of years if not decades.

The differences among the rates of change of individual technologies not only complicate the acquisition of large-scale systems directly, but also increase the difference between the acquisition time scale—typically measured in years and decades—and the changing desires and needs of the user—often influenced by the availability of faster, more capable technologies. Where users can readily acquire technology, particularly information technology, they will do so and home-grow their own systems. At one level, this may be seen as a beneficial stop-gap measure to bridge the interval until the system of record is formally fielded. However, these locally developed solutions gain proponents and thus longevity, clearly exacerbating the problem of finding common solutions across the entire user community. Government and industry have defined numerous acquisition approaches to deal with this mismatch, including evolutionary acquisition[9] and rapid prototyping, and continue to explore the most effective approaches of getting needed functionality to users in as short a time as possible.

5.2.5 Test and Evaluation Challenges and Constraints

Today's government test and evaluation infrastructure must not only evaluate the performance of complex systems against their stated requirements, but also evaluate them in a representative environment. As individual systems define greater numbers of discrete interfaces and as programs are established to engineer and acquire integrated mega-systems, the test and evaluation challenge increases in scope and complexity. Existing test facilities, tools, and methods developed to deal with individual, albeit highly complicated, weapon platforms must be extended to address multi-system and, indeed, multi-service interactions. Constraints on the timely availability of all critical components and on the resources required to move them to a common physical facility have encouraged the DoD to develop an infrastructure to link mega-system components that may be widely distributed physically.

Other challenges involved in testing and evaluating mega-systems stem from the sheer number of interactions and the resultant difficulty of identifying the root cause of failures. Localized phenomena in one part of the mega-system may propagate globally. Failures outside the mega-system may affect performance within its boundaries. Detecting, isolating, and tracing such behavior in a highly complex system is clearly difficult.

Still another difficulty arises from the lack of formal, testable requirements for many mega-systems. Often, formal requirements are stated for the component systems but not for the whole. In other cases, the requirements for the whole are either overly ambitious or fail to capture some new capability that has emerged after the requirements were documented and approved.

5.2.6 Vulnerabilities: Expected and Unexpected Negative Effects

At the heart of all complex mega-systems lies the information infrastructure, which enables the exchange of information among the system's component elements. The increased reliance on the infrastructure brings with it a well-recognized but not fully understood set of vulnerabilities. These are associated not only with cyber threats, such as malicious intrusion, insider threat, and inadvertent errors, but also with the vulnerabilities arising from the inherent complexity of its interactions. We have mentioned some of these. Local failures, whether inadvertent or deliberately caused, can propagate widely. Examples of cascading failures have been well documented in the electric power industry (Amin, 2000). Complex feedback and feed-forward loops can complicate our ability to understand and thus trace the root cause of problems. Many of these vulnerabilities remain invisible until the mega-system is actually constructed and operational; some of them will then emerge over time. These vulnerabilities are inherent in the nature of the interactions and can neither be completely prevented nor resolved through independent modification of the individual components.

Faults in electric lines in Oregon resulted in excess load … which led to the tripping of generators … which led to the separation of the North–South Pacific Intertie near the California–Oregon border … which led to islanding and blackouts in 11 U.S. states and two Canadian provinces (Amin, 2000).

5.3 Troubled Large-Scale Systems

Considerable evidence demonstrates that not all major engineering efforts succeed. In fact, recent years have witnessed a number of spectacular failures. These have, of course, included physical systems (e.g., bridges, tunnels, and spacecraft) and have made front-page news. In addition, there is a long and growing record of software-intensive system failures. Some of these have also made headlines. Many other projects, even if they have not "failed," are seriously over cost and behind schedule.

In 1994, The Standish Group published the first of a series of reports, updated every 2 years since then, entitled the *Chaos Reports*.[10] These reports focus on the state of software projects in the United States and encompass both commercial developments and projects for federal, state, and local governments. In 1994,

- 16% of projects for which data was collected were considered successful; that is, they were completed on time, on budget, and provided all the features and functions initially specified
- 31% were failures; they were canceled at some point during the development process

■ 53% were considered impaired or challenged in that they were completed and operational but over budget, behind schedule, and with fewer features and functions than initially specified

Since then, The Standish Group has reported a continuing improvement in these statistics as well as a continued decline in the average percentage of cost and schedule overruns. By 2006, 35% of projects were deemed successful, 19% were outright failures, and 46% were impaired. These figures certainly do not indicate a complete reversal, but they clearly show a marked improvement over the data reported in 1994. However, results presented in the 2009 report showed a reversal in the improving success rates, with 32% of projects succeeding, 44% deemed challenged, and 24% considered to have failed.

Many sources cite examples of failed information technology projects (Saltzer, 1999, 2004; Charette, 2005; CSTB, 2000). Table 5.3 provides some examples. It is not intended to be an exhaustive list, but rather to illustrate the range of projects that have been canceled after years of effort and considerable expenditures. In some cases, these failures were spectacular and became the subjects of front-page headlines and Congressional hearings.

What caused these projects to fail? Analyses cite many causes, including unrealistic goals, inaccurate estimates, and poor or inappropriate management practices. One could also argue that their failure resulted from the sheer complexity of the functional demands, coupled with a linear, top-down approach inadequate to deal with that complexity. Saltzer (1999, 2004) identifies a number of recurring problems. First, these projects were often efforts to replace existing operational systems with new systems that were overly laden with new features and that relied on new and, in some cases, unproven technologies ("second-system effect"[11]). In some cases, the systems behaved according to a different set of rules as they expanded (incommensurate scaling): features that functioned well in isolation or on a small scale did not work as expected when they were scaled up. Finally, people who were aware of these issues refused to acknowledge them (what Saltzer refers to as the "bad-news diode") or, if they did acknowledge them, those in management who could have acted on their reports, chose not to regard them.

Saltzer (1999, 2004) argues that the solution for these large, complex software projects is to adopt an incremental and iterative approach that focuses on

■ Getting something simple working soon
■ Working on one new problem at a time
■ Identifying ways to find flaws early
■ Using iteration-friendly design

He contrasts this iterative approach, termed "empiricism," with the "rationalist" approach of the top-down plan-specify-build approach of the Waterfall Model

Table 5.3   Examples of Canceled Large-Scale Engineering Projects

Project	Year	Outcome	Cost (U.S.$)
Hudson's Bay Co. (Canada)	2005	Problem with inventory control system led to revenue loss	$33.3m
U.K. Inland Revenue	2004–2005	Software errors contributed to tax-credit overpayment	$3.45b
Avis Europe PLC (U.K.)	2004	Enterprise resource planning system canceled	$54.5m
AT&T Wireless	2003–2004	Customer relations management (CRM) upgrade problems led to revenue loss	$100m
Sydney Water Corp. (Australia)	2002	Billing system canceled	$33.2m
Federal Bureau of Investigation Virtual Case File	2000–2005	Replacement for paper file system canceled	$100–170m
Bureau of Land Management automated land and mineral records system	1999	Canceled after 15 years of effort	$411m
Internal Revenue Service tax modernization	1997	Canceled after 8 years of work	$4b
State of California vehicle registration systems	1994	Never deployed	$44m
Denver International Airport Baggage System	1994	Problems with routing baggage delayed airport opening by 11 months	$1m/day
U.S. Federal Aviation Administration Advanced Automation System	1994	Terminated after 12 years	$6b

Table 5.3 (continued) **Examples of Canceled Large-Scale Engineering Projects**

Project	Year	Outcome	Cost (U.S.$)
CONFIRM reservation systems (Hilton, Marriott, Budget Rent-a-Car, American Airlines)	1992	Canceled after 3.5 years of development	$125m
London Ambulance Service (U.K.)	1990; 1992	Dispatch system canceled in 1990; second attempt started in 1991 and abandoned after deployment with 20 lives lost in 2 days	$11.25m; $15m

(Figure 5.3). The terms "rationalism" and "empiricism" to contrast these two approaches were first used by Brooks in a 1993 lecture.[12]

Saltzer and others have questioned whether traditional top-down systems engineering approaches that concentrate on decomposition and partitioning apply to the development of these large-scale, complex systems. They also question the extent to which such traditional systems engineering practices can continue to scale up, their ability to address systems with complex (nonlinear) behavior, and their extensibility to efforts for which there is no single identified project. In other words, they question whether these techniques apply to the kinds of problems that are encountered in mega-systems.

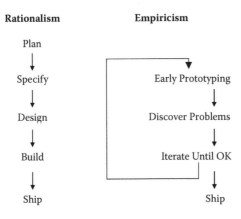

Figure 5.3 Rationalism versus empiricism. (*Source:* Copyright by Jerome H. Saltzer. Used with permission.)

5.4 Levels of Systems Engineering

Traditionally in U.S. defense programs, systems engineering has been a function that resides within a program, with the systems engineer reporting to the program manager. As such, systems engineers, working in support of the program manager, are responsible for delivering a desired capability, sometimes by integrating already-existing or developmental systems and at other times by developing such a capability from scratch. They must design and implement the total system with particular emphasis on translating the users' needs into a viable design. In particular, systems engineers are responsible for generating and analyzing alternative designs and for coordinating and controlling the various engineering tasks involved in implementing the system.

In cases where a mega-system results from the integration of separately managed systems, the systems engineering function continues, but with some significant differences. First, there is rarely a single program structure that encompasses the full set of systems that could be involved. In those cases, system engineers do not work within a single program but rather across multiple programs. Therefore, they are rarely in a position to direct, but instead must be able to influence the various programs and their technical directions. Second, in these circumstances, systems engineers typically direct their attention across the various systems rather than addressing the totality of the interactions. The emphasis is on the interfaces among and the interactions between the several individual systems. We refer to this as "end-to-end engineering" to connote that the objective of this activity is to ensure that all the systems that support a particular operational sequence of activities can, in fact, work together from the beginning to the end of the sequence.

A systems engineer can also support the strategic aspects of shaping an enterprise or even an extended enterprise. These aspects focus on developing strategies, formulating policies, and establishing processes and governance structures. In such cases, the role of the systems engineer is to provide enterprisewide technical guidance, help in developing and drafting related guidelines, and conduct independent assessments in support of the managers of the enterprise. Figure 5.4 highlights these three levels of systems engineering: intra-program, end-to-end, and enterprise-wide.

5.5 Enterprise Systems Engineering Profiler

The Enterprise Systems Engineering Profiler introduced in Figure 5.5 builds on and elaborates the concepts introduced in both the basic framework and the levels of systems engineering. It is intended as a first step toward the development of a *self-assessment tool* that can help the systems engineer understand the nature and context of the system of interest. It is also intended as a *situational model* that can help systems engineers select and adapt the processes, tools, and techniques most applicable to the particular system problem and context.

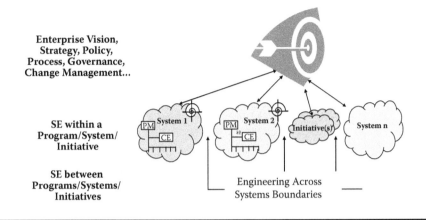

Figure 5.4   Three levels of systems engineering.

5.5.1 Quadrants and Dimensions

The Profiler is organized into four quadrants and three rings. The quadrants describe different dimensions of the broader context in which the system or mega-system will be developed and will operate and evolve. Three of these map directly to the dimensions introduced previously. The fourth quadrant introduces aspects of the implementation or acquisition environment. Each of these four quadrants is, in turn, further decomposed into two related dimensions.

Reading clockwise, the first quadrant addresses the *strategic context*. Here we focus on the dimensions related to the stability of the mission environment and the scope and breadth of the intended system. Requirements for systems that will operate in a stable environment are expected to change more slowly than those for systems that will operate in environments that are themselves changing. More narrowly focused efforts address a single function. As they broaden, they can be expected to address an enterprise or, in some instances, an extended enterprise.

The second quadrant—the *implementation context*—highlights differences in the scale of the effort—the extent to which the program is expected to support a similar community of interest or to span multiple such communities—as well as its structure. This context can range, at its simplest, from a single program established to implement a single system to the obviously more complicated activities associated with multiple programs organized to implement multiple, although related, systems. Note that the acquisition context was not specifically addressed in the framework presented in Chapter 4 but is now included in the Systems Engineering Profiler.

The third quadrant is the *stakeholder context* and directly maps to the decision-making vector of the basic framework. In this model, we have differentiated two aspects of stakeholder involvement: (1) the extent to which stakeholders agree with the goals and objectives of the effort and (2) the extent to which stakeholder relation-

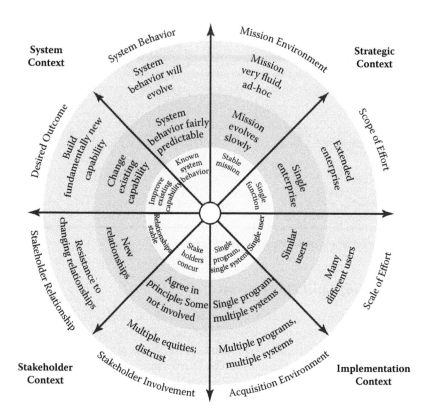

Figure 5.5 Enterprise Systems Engineering Profiler. (From Stevens, R., 2008. *Profiling Complex Systems*, Proceedings of the IEEE International Systems Conference, Montreal, Canada, April 2008. With permission.)

ships change. It is not only the changing relationships that shape the environment, but also the extent to which stakeholders accede to or resist such changes.

The fourth quadrant is the *systems context*. Here we focus on the expected outcome of the effort as well as on the behavior of the system itself. The expected outcome can range from modest improvements in an existing capability to, at the other extreme, the development of a fundamentally new capability. The behavior of the system, described primarily in terms of its predictability, is closely related to the expected outcome. Efforts directed toward improving an existing capability are more likely to demonstrate predictable behavior, while those focused on developing a novel capability are likely to result in behavior that is less predictable and more likely to evolve.

The maturity of the technologies that will be used is a contributing factor in this quadrant. The performance and interactions of technologies at the state of the practice should be well understood and therefore be considerably more predictable. In contrast, the behavior and realized performance of technologies in development or still being explored are obviously far more difficult to predict.

5.5.2 Concentric Rings

As in the basic framework, the concentric rings reflect increasing complexity, uncertainty, and variability as one moves outward from the origin. The innermost band reflects the domain of traditional program management and traditional systems engineering, or at least the idealized state, in which the manager and the systems engineer operate inside the program. Here, the effort is most often characterized by well-bounded problems, stable requirements, mature technologies, and predictable behavior. In effect, this is the realm of "tame" problems and linear approaches to resolving them.

The middle band can be considered the transitional domain. This is the region of end-to-end systems engineering in which the systems engineer primarily works across system and program boundaries. Here, the engineer is more likely to exercise influence than direct control.

The outermost band, which we have termed the "messy frontier," highlights situations where managers and systems engineers must deal with a highly fluid environment, distributed developments often lacking a global blueprint, and multiple stakeholders with independent, and sometimes conflicting equities and systems whose desired functionality and technical behavior are expected to evolve over time. It is the environment of multiple users and multiple stakeholders. As discussed previously, this is the region of uncertainty, unpredictability, and diversity. It is also the region of "wicked" problems—problems that do not lend themselves to the traditional centralized, top-down design and development approaches described above.

As one moves outward along these concentric rings, one encounters fundamental differences in the extent to which the systems engineer can direct change. That, however, does not mean that he or she cannot effect it. In the innermost ring, the region of what we have termed "well-bounded" systems and the province of traditional systems engineering, the systems engineer does have some measure of technical control over the behavior of the systems and management control over all the component elements. As one moves outward, the engineering, development, acquisition, and evolution of these mega-systems take place in the absence of familiar control mechanisms.

The transition from a well-bounded system to a complex mega-system is not a matter of merely scaling up from the well-bounded system. Instead, it involves a significant shift in perspective, in approach, and in the applicability of tools and techniques. The techniques that have emerged to engineer well-bounded systems are predicated on the essential linearity of the systems. Consequently, these techniques may not apply to those aspects of the behavior of mega-systems that are emergent and therefore not predictable. Similarly, the management techniques that work in a unitary environment may not work in a pluralistic one.

This, then, is perhaps the key challenge in engineering and acquiring mega-systems: to develop large-scale, complex mega-systems and then continue to manage their evolution when there is no authority to impose conformity from above. Instead, evolution takes place through the purposeful, deliberate, and cooperative (or in some cases, competitive) actions of the system's constituent elements.

5.5.3 Using the Enterprise Systems Engineering Profiler

This emerging Enterprise Systems Engineering Profiler has two uses. As a self-assessment and a diagnostic tool, it can be used to highlight the nature and context of the system of interest along a number of different dimensions. As a situational model, it can help the management and engineering team select the set of tools that best fits the nature of the problem at hand. In some instances, traditional systems engineering and software engineering processes will continue to be the most appropriate approaches. In other cases, the processes, methodologies, and tools must accommodate the flexibility and adaptability required by the sheer scale, complexity, and heterogeneity of these mega-systems.

In practical terms, the manager or engineer can use this framework to map the system or mega-system of interest, creating a spider chart or polar diagram of the system's context (Figure 5.6).

It should be noted that this map may, in fact, change over time as a result not only of deliberate actions by the management and systems engineering team, but also of events that fall outside their control. For example, the nature of the mission environment may alter, starting out as relatively stable but shifting to one that is more dynamic. Similarly, initial resistance among selected stakeholders may weaken over time and, through the deliberate actions of the program, change to concurrence and support. The opposite may also happen: Stakeholders who initially agree may discover that they have substantial differences over time.

A situational model can help the systems engineer select and adapt the best processes, tools, and techniques on the basis of the system's nature and context. Underlying the very notion of a situational model is the premise that different processes, tools, and techniques apply in different situations. The challenge is to understand the situation sufficiently well to select the most appropriate ones and to adapt the tools as the situation warrants.

An important caveat is warranted here. This Profiler is only the starting point of such a self-assessment tool and situational model rather than a final version. A rich dialogue is required to refine it, and considerably more experience is needed before we can confidently link specific systems engineering practices to particular situations. This is, in fact, a fruitful area of research in the systems engineering community.

Following the two case studies presented in Chapters 7 and 8, we will use the Systems Engineering Profiler to synthesize and summarize the particulars of that case study. The resulting profiles are, in fact, quite different, reflecting the particular circumstances of each case study. In Chapter 10, the Systems Engineering Profiler

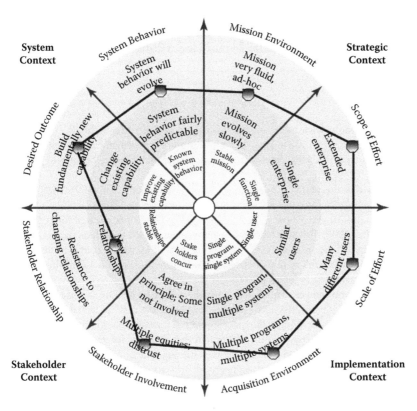

Figure 5.6 Using the Systems Engineering Profiler to map a system or mega-system. (From Stevens, R., 2008. *Profiling Complex Systems*, Proceedings of the IEEE International Systems Conference, Montreal, Canada, April 2008. With permission.)

is repeated, this time as a framework to highlight the emerging tenets associated with engineering mega-systems.

5.6 Summary

This Section provided a highly abbreviated discussion of the nature of large-scale, complex mega-systems and the challenges they present to systems engineers. Both in the commercial world and in government, we value the benefits that accrue from the interactions between disparate systems and the resulting collaborations between previously separate communities. As we set out to deliberately engineer such systems, we find ourselves at the intersection of what we have learned in practice about the engineering of well-behaved linear systems and what some suggest is necessary to develop far more complex, and certainly much less well-behaved, mega-systems.

Endnotes

1. Available online at www.incose.org/whatis.html (accessed 1 March 2008).
2. Software engineering is defined as the "application of a systematic, disciplined, quantifiable approach to the development, operation, and maintenance of software (IEEE, 1990).
3. There is a rich literature on adaptive software development. For information about Scrum, see http://www.controlchaos.com/about/ (accessed 1 March 2008).
4. An initial draft of the *System of Systems Engineering Guide* was released for comment in December 2006. This was followed by pilot phase, which included interviews with practitioners of system-of-systems engineering as well as with research teams active in the field. The results of this pilot phase are reflected in Version 1.0 of this Guide, released for comment in January 2007.
5. Personnel screening and credentialing are formally defined by the DHS as "assessing a person to determine identify, risk and suitability for a benefit, access, credential or privilege."
6. The Western Hemisphere Travel Initiative requires all travelers, when entering the United States, to present a passport or other document that denotes identity and citizenship.
7. The REAL ID Act of 2005 requires people entering federal buildings, boarding airplanes, or opening bank accounts to present identification, including state-issued driver's licenses, that has met certain security and authentication standards.
8. Consolidated screening systems could also have unintended consequences for decision makers such as local law enforcement or immigration officials. For example, if some stopped for a routine traffic violation is found to also be in violation of immigration laws, it would lead to an increased local work load and costs. Similarly, additional criminal information being presented to immigration officials adds additional complexity and time for admittance decisions for non-U.S. citizens.
9. Evolutionary acquisition, also known as spiral development (see Chapter 5, Section 5.1.2), "allows more rapid deployment of systems and provides a process for incremental upgrading of fielded systems.... [It] permits the addition of new capabilities to a system as the underlying technologies evolve without this being viewed as 'requirements creep.'" It is one of three primary thrusts of Chairman of the Joint Chiefs of Staff Instruction 3170.01A, *Requirements Generation System* (Washington, DC: Office of the Chairman of the Joint Chiefs of Staff, August 10, 1999).
10. See The Standish Group at: http://www.standishgroup.com.
11. In computing, the second system effect refers to the tendency to design the successor to a relatively small and successful system as large and feature laden. The term was first used by Frederick Brooks in *The Mythical Man-Month: Essays on Software Engineering, Anniversary Edition*. Boston, MA: Addison-Wesley.
12. The terms "rationalism" and "empiricism" were used by Frederick Brooks in a keynote lecture entitled "Language Design as Design," delivered at the April 1993 Second ACM SIGPLAN History of Programming Languages Conference,Enterprise Systems Engineering Profiler.

CASE STUDIES IN ENGINEERING MEGA-SYSTEMS

Chapter 6

Introduction to Mega-System Case Studies

This chapter introduces two real-world case studies (see Chapters 7 and 8) to explore the practical aspects of engineering large-scale, complex systems that cross functional and organizational boundaries.

6.1 A Note about Case Studies

A case study is a method for learning about a complex topic by considering the topic as a whole and setting it in its broader context. A case study provides a longitudinal description and analysis of a project, a business, or an industry, and chronicles the events that happened, the perspectives of the different stakeholders, the processes followed, the decisions made, and the outcomes achieved.

The fields of law, government, medicine, business,[1] and sociology have a long-standing and rich tradition of using case studies as teaching tools. More recently, we have seen a growing interest in developing case studies in the field of engineering, particularly systems engineering. For example, Hughes's book, entitled *Rescuing Prometheus: Four Monumental Projects that Changed the Modern World,*[2] presents four large engineering projects of the twentieth century as a way of describing the emergence, and the limits, of systems engineering during that time.

Friedman and Sage (2004) offer a more structured approach to systems engineering case studies as a way of illustrating systems engineering concepts. The Friedman-Sage Framework, as it is called, is presented as a matrix of nine systems engineering concept areas[3] and three responsibility domains. These domains differentiate between government, contractor, and shared government–contractor responsibilities. Friedman and Sage suggest an approach of developing case studies to illustrate different cells in the resulting 9×3 matrix.

We are also beginning to see the development of repositories for such case studies. The Air Force Center for Systems Engineering, part of the Air Force Institute of Technology (AFIT), has sponsored the development of several case studies[4] using the Friedman-Sage Framework and posts them on online. Other academic institutions, such as the Harvard Business School, are also writing case studies and making them available not only to their own students, but also more broadly to systems engineering practitioners.

6.2 Approach to Mega-System Case Studies

The case studies presented in this section are intended to tell a story: how each of these efforts tackled the engineering of a particular large-scale, cross-boundary project. The studies are intended to describe rather than to assess or critique. In presenting these studies, we hope to identify patterns and glean insights about what techniques seem to work and the circumstances in which they are effective.

Unlike traditional case studies, in which the effort presented is complete and outcomes can be examined, these case studies examine activities that were, at the time of writing this book, still in process. Consequently, we can describe the practices and techniques being implemented, but we cannot always confidently predict whether they will, in fact, achieve the desired outcomes.

The case studies presented here were selected on the basis of several criteria:

- They address efforts that approximate our definition of a mega-system.
- They invoke the engineering of mega-systems. That is, there is a deliberate, managed effort to achieve a desired outcome.
- Information about the project events and engineering approaches is readily available through a combination of interviews, project documentation, and published material.

Both of the programs described as case studies depend, in whole or in part, on information technologies to achieve their objectives. The SIAP is a DoD-wide effort to develop a common air picture that all military units in the battle zone could use to detect and track all airborne objects and distinguish between friend and foe. The second case study comes from the commercial sector. It focuses on the development and deployment of RFID technologies to identify and track items throughout the global supply chain.

These case studies also illustrate two different kinds of mega-systems. SIAP is an example of a composed mega-system, in that the effort is intended to fix known inconsistencies in the way that different systems detect, identify, and track air objects. RFID explores an approach to what we have termed designed mega-systems. Rather than integrating already-developed components, this effort is directed at engineering and developing fundamentally new capabilities that span functional or organizational boundaries. In the RFID case, the boundaries stem from corporate concerns over protecting proprietary data; the desire to maintain competitive advantage keeps suppliers and their retailer customers from sharing information about items in the supply chain.

Neither of these projects is typical within its larger organizational context. SIAP is not a traditional DoD acquisition program—although it may be restructured to become one—but a systems engineering activity involving multiple acquisition programs. RFID is not an effort to produce a particular system, but rather an attempt to develop a set of standards that could be used by multiple vendors. What they do have in common is that each effort is designing a capability that is intended to span multiple, independent users.

Both case studies begin with a discussion of the background, focusing on the circumstances and events that led to the establishment of the project. We then show how the project evolved over time. Throughout, we highlight the engineering and management processes implemented and, where these have changed over the course of time, identify the reasons for such changes. At the conclusion of each case study, we use the Enterprise Systems Engineering Profiler presented in Chapter 5 as a means to synthesize the circumstances of the case.

At the end of Section III there is a brief discussion of the similarities and differences between the two cases. In so doing, we seek to highlight topics relevant to the engineering of mega-systems.

Endnotes

1. The case study method was pioneered by the Harvard Business School in the 1920s as a way to engage students in exploring real-world problems and decisions.

2. The projects described are the Semi-Automated Ground Environment, a computer- and radar-based air defense system; the Atlas project, which developed the first U.S. intercontinental ballistic missile; the Boston Central Artery/Tunnel Project, also known as the "Big Dig," which is building a complex system of tunnels and bridges to move traffic through downtown Boston; and the ARPANET, an early interactive computer system that is recognized as the progenitor of the Internet.

3. The nine concept areas are Requirements Definition and Management; Systems Architecture Development; System, Subsystem Design; Validation and Verification; Risk Management; Systems Integration and Interfaces; Life Cycle Support; Deployment and Post Deployment; and System and Program Management. The first six represent the faces of the systems engineering life cycle, while the latter three encompass process and systems management support.

4. The case studies can be found online at: www.afit.edu/cse/cases. They cover the B-2 stealth bomber, the C-5A and C-5B aircraft, the F-111 fighter-attack aircraft, the Global Positioning System (GPS), the Hubble Space Telescope, and the Theater Battle Management Core System (TBMCS). While all these case studies describe technical and engineering challenges, five of them are about essentially well-bounded projects and one focuses on an information-based system.

Chapter 7

Single Integrated Air Picture[1]

A coherent picture of the battlespace is a critical enabler for joint warfare. When such a picture depicts and tracks the location of all aerospace objects and distinguishes friend from foe, it is called the Single Integrated Air Picture (or SIAP for short). The primary network used by all the military Services today to create such a picture is called Link 16.[2] Link 16 operates on two advanced radio transceiver terminals: the Joint Tactical Information Distribution System (JTIDS) and the Multifunction Information Distribution System (MIDS).

While a formal standard defines the messages that will be passed on this network,[3] the standard is lengthy, complex, and allows individual weapon system platforms (encompassing aircraft as well as ground- or sea-based air defense systems) to implement it differently. It includes options, leaves interpretations open to individual platforms, and does not address platform-specific issues such as databases or displays. Thus, different platforms may have interpreted the standard differently, implemented different features, or opted not to implement a particular set of features. The result: different pictures in different platforms, operator confusion, and—all too often the risk of fratricide. To address this well-recognized problem, the DoD established a formal SIAP program in 1998.

7.1 Motivation: Moving from Independent Systems to a Theater-Wide Integrated Capability

Traditionally, each military Service has developed its own weapon systems with the expectation that these would be used independently; that is, that each Service would be able to direct the use of its own weapons. However, over the past decade or so, it has become clear that, rather than operating independently, Service-developed and -operated systems would have to function as part of a joint integrated capability.[4] Sometimes, as in the case of air and missile defense, that capability would be theater-wide and under the management of the regional combatant commander.[5]

The 2010 Joint Theater Air and Missile Defense Concept, prepared by the Joint Theater Air and Missile Defense Organization (JTAMDO) in 1996, articulated this transition from independent, geographically based air and missile defense operations to a theater-wide, integrated capability that leverages long-range weapons and overlapping weapon engagement zones.

The goal of this vision is to build a theater-wide, integrated, joint force capable of destroying TAMD (theater air and missile defense) targets at the time and place of the commander's choosing—on the ground, before and after launch, or in flight, in support of the Defense Department's TAMD objectives. To realize this goal, joint employment concepts achieved through interoperability are needed to realize the full benefits of long-range weapons and overlapping weapon engagement zones. Current doctrine; concepts; and tactics, techniques, and procedures (TTPs) based on individual weapon systems and deconfliction rules must move toward an integrated, collaborative approach centered on joint planning and engagement. Today's isolated individual weapon system engagement zones that restrict warfighting options must evolve into theater-wide joint engagement zones (JEZs) for attack operations and active defense (JTAMDO, 1996).

Key to the accomplishment of such an integrated capability is a coherent state of situation awareness among all military units operating in the battle zone to detect and track all objects (including ballistic missiles) and to identify them as friend or foe. This shared situation awareness forms the basis for a SIAP and is defined as "the product of fused, near-real-time and real-time data from multiple sensors to allow development of common, continuous, and unambiguous tracks of all airborne objects in the surveillance area" (JTAMDO, 1998). The SIAP is being developed from an integrated capability that merges data from multiple sensors and will provide all system operators with a consistent and complete set of information about each airborne entity. This capability is viewed as a critical operational requirement.

Problems with achieving such a SIAP have been widespread, chronic, and well documented. They have been noted in real-world operations, in exercises, and in periodic test and evaluation events. These problems include erratic tracking, dual or multiple track designations, inconsistent location information, and incorrect identification, among others. They have resulted in a confused air picture where some airborne objects do not show up on the display, where they show up but are misidentified as either friendly or hostile, or where the display shows two aircraft where there is only one. Furthermore, different platforms may display different results. Such confusion has translated into, at a minimum, increased operator workload, and, at worst, cases of "friendly fire."

The challenge for the SIAP did not lie in finding technical solutions to these problems. Root causes had been identified and understood. In fact, many of the root causes revolved around differing interpretations and implementations of the Joint Data Network standards.[6]

The challenge was not to find the technical solutions, but to come to *jointly agreed-to* solutions and then coordinate the implementation of these solutions by multiple, separate programs.

Instead, the challenge was one of coming to *jointly agreed-to* solutions and then coordinating the implementations of these solutions.

7.2 Standing up a System Engineering Organization[7]

To overcome inadequate cross-Service engineering and different implementation of standards, and in anticipation of the need to develop and field the desired SIAP capability, the DoD established a special organization—the Single Integrated Air Picture System Engineering Task Force (SIAP SETF)—in October 2000. As described in the Undersecretary of Defense memorandum establishing the SETF (Gansler, 2000), its objectives "include performing the system engineering needed to fix problems in the existing Joint Data Network and to guide development toward a future SIAP capability."

The mission of the SIAP SETF, as stated in its charter, was to identify the most efficient means to achieve a SIAP that satisfies warfighter needs. More specifically, the Task Force was charged with

- Implementing a disciplined system engineering process
- Making recommendations on the most cost-effective solutions that lead to measurable improvements in warfighting capability
- Providing technical expertise to aid in SIAP requirements development
- Developing system and technical [architectural] views for the SIAP component of the TAMD integrated architecture.

The SETF was structured as a small (approximately 30 full-time staff) organization, staffed with personnel supplied by each of the military Services. A "virtual" staff of subject matter experts from the various Services, funded by the SETF, would supplement the "core" staff.

The charter defined the responsibilities of the SIAP Acquisition Executive, the SIAP Oversight Council, and the SIAP System Engineer, and the related roles of the Services and agencies. Of note, the SIAP Acquisition Executive came from the Army (the Army Acquisition Executive), the SIAP Technical Director from the Navy, and the day-to-day manager from the Air Force. These latter two jobs were defined as full-time dedicated positions. While the SETF was assigned responsibility for centralizing the SIAP system engineering functions and producing service-coordinated recommendations, the individual Services and agencies retained responsibility for execution.

The SIAP SETF was not a typical acquisition office, but was designed as a special system engineering organization that took direction from its own management structure. Oversight came through two channels. The first was through the established operational requirements process within the Joint Staff, where approval authority for both validation and prioritization of recommended solutions to existing systems is vested in the Joint Requirements Oversight Council (JROC).[8] The second was through the SIAP Oversight Council, a three-star-level body with representatives from each of the military Services. The SIAP Oversight Council served as an advisory body to the SIAP Acquisition Executive and was the group to which unresolved system engineering matters could be elevated. This second channel was intended to focus on coordinating the implementation of approved solutions.

In addition to these two direction and reporting channels, the SIAP SETF had to coordinate with several other organizations with similar responsibilities for resolving interoperability problems affecting air warfare. The SETF established a close working relationship with the Joint Forces Command (JFCOM) on issues related to priorities among problems to be solved as well as on procedures to mitigate operational ambiguities. The SETF also built close alliances with JTAMDO, responsible for the operational air and missile defense architecture; with the Ballistic Missile Defense Office (now the Missile Defense Agency (MDA)), responsible for the system engineering for the air and missile defense architecture; and with the Defense Information Systems Agency (DISA) on process issues related to Joint Interoperability of Tactical Command and Control Systems (JINTACCS) and coordination with allies.

7.3 SIAP System Engineering Process

The SIAP SETF charter directed it to implement a disciplined system engineering process. That initial system engineering process involved four steps: (1) requirements analysis, (2) functional analysis, (3) implementation analysis, and (4) phased implementation (see Figure 7.1). These steps were adapted from traditional system engineering principles and practices. The formal products expected from this process were system baselines and Block Improvement Plans to define the evolutionary deployment of incremental SIAP capabilities. As a by-product, the process was intended to help develop approaches to reach agreement on Link 16 implementations, test these implementations, and determine their warfighting contributions, and to resolve disagreements among the different Services.

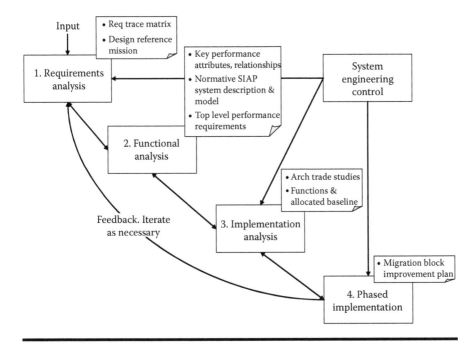

Figure 7.1   SIAP system engineering process.

7.3.1 Initial Strategy: Incremental Blocks

The SETF structured the effort as a series of incremental blocks. For each block, the SETF would identify the engineering specifications; document the supporting rationale, including test results and analysis; and estimate the resources required to implement the recommendations across the various systems. Implementation, as noted above, was the responsibility of each of the Services.

7.3.1.1 Block 0: Demonstration of SIAP System Engineering Process

The initial focus of the Task Force, referred to as Block 0, was threefold. First, Block 0 was to establish a disciplined engineering process by which to develop and integrate SIAP recommendations. Second, it was to develop recommendations for near-term fixes to known Link 16 problems. Finally, the SETF was to lead the development of the system and technical views of the SIAP architecture as input into the broader TAMD integrated architecture.[9]

Block 0 activities focused on four significant link interoperability issues selected and subsequently endorsed by the JROC because of their impact on joint theater operations, their applicability across all the military Services, and their engineering maturity. These issues were correlation and decorrelation, identification taxonomy and symbology, identification conflict resolution matrix, and formation tracking assessment. The first three already had agreed-to solutions in the form of proposed changes to the Link 16 standards, and two of those three had already been approved as Interface Change Proposals (ICPs) to the military standard. These four items were intended to serve as the initial demonstration of the SIAP system engineering process.

Given this situation, the key engineering activities during this phase centered on analyzing the impacts that implementing these solutions would have on each system affected. Block 0 activities also included development of an initial baseline architecture.

To accomplish these tasks, the SIAP SETF created several system engineering teams consisting of core members of the SETF staff as well as representatives from each of the military Services and the Ballistic Missile Defense Organization (BMDO), now the Missile Defense Agency (MDA). The SIAP System Engineer formally signed out requests for information from the SETF to the Services; SIAP funded the Services to provide data to and participate in these engineering teams. Nevertheless, responses were often late and incomplete. Funding shortfalls and restrictions on access to necessary engineering detail hampered the ability of the SIAP SETF to accomplish all the tasks set forth at the outset. Despite this, the Task Force was able to complete an "as-is"[10] architecture for Link 16 that incorporated the 12 systems selected for Block 0 implementation. Further, the SETF was able to begin the design for a solution to the fourth item.

The individual Services constantly emphasized the need to define the operational utility (or benefit) of each of the proposed fixes relative to the resources that would have to be expended to implement it. In response, the Task Force had planned to institute a structured analysis approach and had developed an Integrated Assessment Plan to guide the analysis. However, the proposed activities were again severely curtailed by time and resource constraints. In fact, by the end of 2001, the Integrated Assessment Plan existed only in draft form and the assessment conducted for this initial phase relied on a combination of previous assessment activities and modeling and simulation "check-out" results. Problems with validation and verification of the models used to generate such operational results persisted; validation proved difficult and approvals were not forthcoming. Despite these acknowledged shortcomings, and with a set of well-documented caveats, the Task Force concluded that the three recommended fixes should be implemented.

Tensions surfaced between the SIAP System Engineering Task Force with its cross-cutting mission and the Services that were responsible for funding and implementing the Task Force's recommendations.

In December 2001, the SIAP SETF briefed the JROC on its recommendations, and the JROC approved the recommendation to implement the three ICPs.

The military Services were responsible for budgeting and implementing the JROC-approved changes and were directed to do so within their existing resources. In virtually all cases, the Services had to allocate this funding within the mission area; that is, they had to fund air defense fixes with air defense dollars rather than allowing diversion of dollars from other accounts or new money. Thus, implementing Block 0 could have an impact on other programs that the Service might view as more important.

Work continued for several months on refining the report documenting the recommendations previously presented to the JROC. However, one of the Services notified the SETF that it could not endorse the final draft, citing two areas of disagreement. The first dealt with the models that had been used to generate the analytic results supporting the recommendations. The concern was that these models were not validated and verified, and therefore the results were not definitive. The second focused on the SETF's recommendation to move to a single standard for Link 16 that would apply to both the United States and NATO, contending that this recommendation was not achievable.

With Block 0, SIAP demonstrated a good-faith effort to develop technical solutions that spanned multi-Service implementations. Technical representatives of the various Services were able to work together to develop common solutions. However, when it came to implementation, the institutional interests of the individual Services dominated. One of the recommendations included in the Decision Support Binder, prepared by the SETF, was: "Require that approval of ICPs be accompanied by a commitment to implement the approved change, including a commitment to support funding of the change (SIAP SETF, 2001)." Directed at the Services, it was intended to link approval of the change proposal with the resources necessary to implement the change. The perceived need to make this recommendation provides a glimpse into some of the tensions that surfaced between the SETF, with its crosscutting engineering mission, and the reality that implementation of its recommendations was reserved to the Services, each one using its own funds and acting in its own interests.

A second area of tension also emerged. The JROC and JFCOM, which served as the warfighter representative, pressed the SETF to demonstrate some near-term impacts.[11] Yet, at the same time, the Services demanded rigor and substantive justification to support recommendations and they reserved the right not to concur on process.

7.3.1.2 Block 1: Taking on Additional Technical Issues for Resolution

Notwithstanding the questionable success of Block 0, the next block started in January 2002 and was scheduled to conclude in September 2003.

Like Block 0, this block initially sought to remedy previously identified problems but, unlike Block 0, it did not begin with a preselected set of issues. Instead, the approach adopted was to identify a set of draft goals, later termed "themes."

To do this, the SETF established an Issues Development Group, consisting of subject matter experts from the military Services and the Missile Defense Agency (MDA). Drawing on their experience and engineering judgment, the group identified a list of four operational benefits (operational themes). Two of them, "Further Reduce Dual Tracks" and "Improve Combat Identification," were selected to leverage Block 0 accomplishments and to fix long-standing problems. The third theme, "Improve Tactical Ballistic Missile Defense Performance," was selected to involve the MDA more directly in the SETF's efforts. The fourth theme, "Improve Data Sharing," was selected primarily to set the foundation for subsequent SIAP efforts. Eventually, the group identified a total of 13 technical issues to address.

> Endorsement of the proposed operational focus provided an operational context in lieu of the traditional statement of requirements.

When the SETF provided the preliminary draft goals to JFCOM, the (then) Deputy Commander in Chief, JFCOM, endorsed the proposed operational focus.[12] This endorsement provided an operational context in lieu of the traditional statement of requirements.

At the start of Block 1, the SETF developed draft statements of work for the Block 1 technical issues and forwarded them to the Services for comment. These task descriptions included both global and issue-specific tasks. The statement also included a preliminary list of the systems likely to be affected by these issue areas. Once the statements of work were finalized, the SETF would deliver them to the Services, accompanied by necessary funding, to perform the specified engineering and analysis tasks.[13]

Block 1 was clearly far more ambitious than Block 0 had been. First, it sought to address 13 separate issues, few of which had preexisting, acceptable solutions. Contrast this to Block 0, which addressed four issues, three of which already had agreed-to solutions. These Block 1 issues also had a significantly broader scope, because they included missile defense in addition to the air defense topics that had been addressed in Block 0.

Second, the Block 1 issues were in various stages of maturity. Some were still in the requirements phase, while others were understood in principle but needed further functional analysis or synthesis. Still others were mature, with the issues well understood.

During calendar year 2002, the SETF placed its primary emphasis on two of the four themes: "Further Reduce Dual Tracks" and "Improve Combat Identification." Although limited progress was made on the other two themes, the SETF considered the efforts in the "Improved Data Sharing" theme effective in laying the groundwork for future SIAP block efforts.

Many of the system engineering activities during this phase centered on the development of Block 1 architectural products. These products, including functional decomposition, system views, and technical views, were considered critical to the ongoing analysis of system implementation and the subsequent determination of warfighting benefits. The SETF used a variety of means, including digital engineering-level modeling and simulation tools,[14] hardware-in-the-loop (HWIL) events,[15] and live events,[16] to analyze Block 1 issues. They conducted root cause analysis to isolate performance discrepancies and defined a number of critical experiments to help in the development of solutions.

Nevertheless, it proved difficult to actually reach closure on solutions. The SIAP SETF tried to use the metrics (called SIAP attributes) that had been defined during Block 0 and further extended in Block 1 as an element in formal decision criteria, but found it difficult to come to agreement on the tools that would be used to determine the extent of improvement in actual SIAP performance and, in particular, the resultant impact on mission outcome.

Notwithstanding approval of the operational focus, an ongoing challenge that the SETF faced was the lack of any formal operational requirements specific to SIAP. The operational concept and operational requirements for SIAP were established in the Joint Theater Air and Missile Defense (JTAMD) mission area, but these were considered too general. More specific and granular requirements were needed to conduct the system engineering necessary to define the objective SIAP.

Consequently, the Task Force, working as part of the broader JTAMD process, developed an operational framework for SIAP. The SETF conducted a requirements analysis, extracting SIAP-relevant requirements from existing documents and mapping them to the SIAP activity model. This yielded a consolidated set of "derived SIAP operational requirements." In addition, the Task Force developed a top-level SIAP operational concept as well as a set of detailed operational assumptions. These operational assumptions were believed necessary to frame what SIAP is and is not, and to describe SIAP operations more specifically. However, these products were not offered up as "requirements," largely because the SIAP SETF was specifically precluded from developing operational architectures. In fact, according to its charter, "The SIAP SE task force will not establish operational requirements, but will provide technical expertise to aid requirements and concept of operations development (SIAP SETF, 2000)."

> In the absence of formal requirements, SIAP developed a set of "derived" requirements, an operational concept and detailed operational assumptions.

By December 2002, when the SETF provided its interim report on Block 1 issues to the JROC, it was able to make some firm recommendations backed by analysis. Other issues were still being defined and still others were in the process of developing potential solutions for test.

In parallel with development of the operational framework for the objective SIAP, the engineering activities also entailed development of initial architecture products for the core SIAP Block 1 systems. The views depicted the currently fielded—the "as-is"—Link 16 architecture incorporating the Block 1 core systems. However, funding shortfalls limited the scope of the efforts and the SETF was only able to deliver a first installment of the architectural design.

7.3.2 Strategy Shift: From Developing Solutions to Building a Model-Driven Architecture

In January 2003, the program experienced a radical change in direction, as the SIAP system engineering effort changed from a block upgrade approach focused on fixing problems with implementation of Link 16 to an architecture-based process intended to define common tactical Battle Management Command and Control (BMC2) functionality.

7.3.2.1 Model-Driven Architecture (MDA)®

While development and analysis of Block 1 issues continued, the emphasis within the SETF significantly changed in January 2003. The SETF members had become convinced that, no matter how well they developed paper standards and specifications, they could not guarantee consistent implementation. As a result, the SIAP SETF initiated an effort to design and develop a "computerized" specification that would implement the desired SIAP capabilities. While the program did not specifically sign up to develop a computer program for Service weapon systems, the tools they used to develop the specification allowed them to do that. Code is a by-product of the model and the tools. To a large extent, three factors drove this shift: (1) the need to reduce costs, (2) the desire to shorten the time to fielding, and (3) the expectation that common and consistent implementation would yield improved operational performance—in effect, pay less and do it faster. To do so, they adopted the Model-Driven Architecture (MDA)® approach.[17]

> Shift away from developing paper standards and specifications to developing a "computerized" specification.

MDA is both a standards framework and an approach to developing application software that is independent of the specific technology (hardware or middleware) that implements it. The implementation technology is referred to as the "platform"; thus, in MDA, the term "platform" refers to the underlying computer technology rather than the weapon system in which the SIAP-developed functionality will be implemented. By separating the fundamental application logic from its underlying platform, the MDA approach promises to enhance software reuse and portability, improve cross-platform interoperability, and significantly reduce the time it takes to migrate to new platforms as they are implemented. In effect, the MDA seemed to offer a new way of writing specifications.

The MDA approach was selected to shift SIAP from a paper specification and associated standards to an executable behavioral model. Paper standards are inherently limited in that they have gaps, overlaps, and conflicts. In fact, the larger and more complex the standard, the more likely it is that it will have these limitations.

In addition, paper standards provide only a static description, whereas the behavioral model was expected to be unambiguous and dynamic. This approach promised the following benefits:

- Create the ability to express the behavior of the distributed system in an industry-standard language, Executable Unified Modeling Language (UML™).[18]
- Create the ability to enforce the behavioral model and to ensure interoperability of all participating systems.
- Allow verification and validation of the integrated architecture.
- Change the configuration management artifact from a paper standard and associated source code to a behavioral model.
- Support the verification and validation of the end product.

7.3.2.2 Integration into Warfighting Platforms

In the business model underpinning both Block 0 and Block 1, each solution was developed separately and then the costs estimated separately for every system affected. The individual programs then had to fund the design and coding for each solution. The estimated costs, summed across each of the programs involved, were deemed too high. The new MDA approach replaced the separate implementation efforts with a single design and coding activity, funded and performed by a team collocated at the SIAP headquarters. However, the costs of integrating the resulting code into the weapon systems would still be the responsibility of the individual Services and their weapon programs.

For SIAP, this common specification is referred to as the Integrated Architecture Behavior Model (IABM) Platform Independent Model (PIM). Thus, the MDA is the approach; the IABM is the executable model that specifies the desired functionality; and the PIM is the definitive model of the application that separates that functionality from the underlying implementation technology.

The IABM is intended to capture the processing logic and encapsulate SIAP requirements and specifications in an executable UML™ tool that is used to generate a behavioral model. Because it reflects the expected SIAP behavior in a model, the IABM is considered both more precise and less subject to varied interpretation by prime contractors than the traditional paper specification. In essence, it is expected to show industry "what 'good' looks like" (SIAP SETF, Undated).

The PIM, as its name implies, is independent of the particular computing environment. To integrate the PIM in a specific military platform, it is necessary to translate it into a Platform Specific Model (PSM). Only then can the resultant model be integrated into the operational platform. After a PIM is developed, it is automatically converted, via machine transformation, into one or more PSMs[19] that take into account the implementing hardware and middleware specific to each system. This tailors the PIM to the computing devices, operating systems, middleware, and communication interfaces associated with a particular warfighting system. From there, the PSM is compiled into a computer program for implementation in the individual systems.

SIAP funded several contractors responsible for engineering the target systems to work the translation from the PIM to the PSM and then to integrate the results into a warfighting system. The underlying expectation is that, because each of these implementations is to be based on the same behavioral model, their individual behaviors will be predictable and consistent across all the target platforms.

A key point is worth noting here. If the integration test results identify any necessary corrections or changes to functionality, they are to be implemented in the PIM itself and not in any of the subsequent instantiations of that functionality. This integration process is performed separately for each of the platforms that integrate the common functionality. Figure 7.2 provides an overview of this integration process.

While the initial focus of the PIM effort was on-track management and combat identification, from the beginning the intent was to expand the scope significantly to address common applications associated with joint tactical battle management, and command and control functionality. In anticipation of this more extensive challenge and in parallel with the initial efforts, SIAP established a team to define required technical behavior, with an emphasis on the coordinated behavior of the "ensemble,"[20] that is, the various SIAP platforms working together.

The goal of this effort was, from the outset, envisioned as a federation of peer systems that, while they may be deployed on a wide range of heterogeneous platforms, all implement the same processing logic.

Each peer is a platform-specific implementation of the same IABM and, as such, could share consistent information such as combat identification. Further, each peer can generate an improved and more complete picture of the battlespace by using information it receives from the other peers (Krikeles et al., 2004).

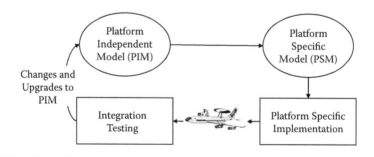

Figure 7.2 SIAP integration process.

7.3.2.3 Distributed Test Environment: Joint Distributed Engineering Plant (JDEP)

Over the past decade, one of the most important lessons learned in analyzing enterprise-level systems is that these systems are difficult to represent entirely within a single simulation. This occurs partly because of the complexity of such systems and partly because it is difficult to gain agreement on how to represent all the elements of the enterprise when multiple organizations own the constituent systems.

At the same time as the SIAP program was being established, the DoD instituted an effort to develop a distributed engineering environment to support system-of-systems engineering and testing. This initiative, known as the Joint Distributed Engineering Plant (JDEP), supports integration and interoperability testing of joint systems of systems. JDEP formally began in September 2000 with an initial focus on JTAMD, to a large extent because this area had a recognized need (Dahmann and Crisp, 2003/2004).[21]

The JDEP paradigm allowed organizations with authoritative simulations of their systems to bring them into the distributed environment so that the system-of-systems can be represented by a system of simulations. The testing environment envisioned both an HWIL environment and a constructive federation environment. This federation was developed, integrated, and tested over several iterations known as JDEP Infrastructure Builds (IBuilds), which occurred simultaneously with the development of the IABM. In essence, it can be viewed as a developmental test harness that IABM developers can use to test versions of the IABM incrementally as they are developed. Thus, early on, the planned system can be tested against a federation of constructive simulations that permits faster-than-real-time analysis. As a component evolves, it can be tested against realistic HWIL systems that run in real-time to gain a better understanding of how the component will perform when working with actual systems.

From the beginning, a close relationship has existed between the JDEP and SIAP system engineering effort. In fact, JDEP theory and practice have evolved largely through interaction with SIAP. Conversely, the application of a distributed simulation environment composed of multiple simulations, each of which focused on representing a different system that contributes to the SIAP, was a compelling argument for SIAP to adopt JDEP as their test and evaluation environment.

In 2001, JDEP executed a proof-of-concept event and identified four sites that cut across the Services and represented key JTAMD systems. The original plan was to link these four sites using the Navy Distributed Engineering Plant battleground testing approach,[22] drive the test with their scenarios, and analyze the results in terms of "capabilities and limitations" by applying Navy metrics and analysis tools. Using seed money from JDEP, the SIAP SETF became the "customer" for the event. As a result, the proof-of-concept event was refined to focus on linking these sites to address specific SIAP issues.

The SETF and JDEP required considerable time and effort to determine how this four-site environment could be used to meet SIAP needs. Assumptions made early in the JDEP planning process did not always meet SETF needs. As SIAP developed its own metrics and analysis tools, the JDEP team had to adapt its approaches to accommodate them. While this initial event did not meet all the needs of the SIAP SETF, it was viewed as a valuable learning experience. In particular, it highlighted the need for a common bridging infrastructure that could be used to link various existing HWIL and simulation environments.[23] As such, it contributed directly to the development of the JDEP as supporting multiple, concurrent federations of HWIL systems and simulations.

Building on the JDEP strategy, process, and framework, the SIAP SETF partnered with JDEP to conduct a series of "pilot" federation events. The first[24] was held in November 2002 at the E-2C System Test and Evaluation Laboratory located at the Naval Air Station, Patuxent River, Maryland. Others were held at the Patriot test facility in Huntsville, Alabama, and the Lockheed Martin AEGIS test facility in New Jersey. Each site was configured to run an HWIL-based event followed by a simulation event.[25] Then, a "combined" event was held in the summer of 2004 to federate these separate sites.

7.3.2.4 Organizational Changes

At the same time as the SIAP effort was evolving toward an MDA approach, the organization responsible for the design, development, and implementation of the SIAP IABM was also changing. In January 2003, the SIAP System Engineering Task Force was transitioned into a Joint SIAP System Engineering Organization (JSSEO) with the objectives of simplifying oversight, establishing clear funding lines within each Service, and being able to dedicate sufficient resources to support the initiative.

The IABM is being developed by a SIAP-led and -funded consortium, which includes the contractors responsible for building the systems into which SIAP capabilities will be integrated.

The JSSEO functioned as a collaborative organization in which a relatively small core engineering and management team was augmented with technical expertise from industry and the academic community (Dutchysyn, 2005). The military Services contributed subject matter experts to participate in the SIAP engineering process. Thus, the IABM was developed by a SIAP-led and -funded consortium consisting of operational and technical subject matter experts as well as developers. Consortium members include representatives from industry, academia, Federally Funded Research and Development Centers, and the various Services, along with contractors who work on the systems targeted to receive SIAP capabilities and the JTAMDO, the organization responsible for setting SIAP requirements. Unlike the earlier phases of SIAP system engineering, in which representatives from different organizations formed a "virtual" team, the members of this consortium were collocated at a central development site.

7.4 Building the IABM

7.4.1 Incremental Configurations

JSSEO members followed an aggressive development approach in which some amount of working code was delivered every several weeks as a time-phased "beta" delivery,[26] referred to as a time box (see Figure 7.3 for an overview). The process, as initially implemented, started with a Requirements Definition Document (RDD) that defined the requirements and architecture for that time box. Next, modifications were made to the incremental IABM architecture and documented in UML. Model implementers generated the IABM using an executable UML case tool.[27] The tool supports an Action Semantic Language (ASL), which, when combined with the UML Model, is sufficient to generate source code (in this case, C++) automatically.

When the time box implementation was completed, the development team assembled a Version Description Document from project and domain notebooks that were also generated by the tool. Initially, 4 weeks were allocated for development, followed by a 1-week installation and check-out phase and a second week for unit testing. From there, the code was placed into configuration management and a copy was burned on a compact disc and passed to the target systems and beta sites for review and check-out in their own labs. The output constituted a portion of the PIM. The time boxes overlapped, with development of the next time box starting up as installation and check-out of the previous one got underway.

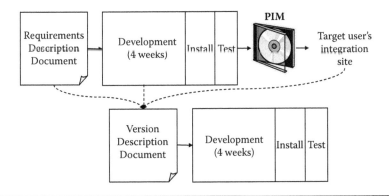

Figure 7.3   Initial SIAP time box process.

Feedback from the users, lessons learned during the development of the time box, and any deferred functionality from that time box were incorporated into the Requirements Description Document for the subsequent time box.

IABM time box testing was conducted at the JSSEO facility in Crystal City, Virginia, using IBuild software that had completed development, integration, and test at MITRE-McLean. The development team was responsible for producing the IABM; from there on, implementation became the responsibility of the individual systems affected and their contractors.

Target system developers have established SIAP-funded integration sites where they can review and load the code. They have also received training to understand the SIAP architecture and artifacts. It should be noted that the product delivered at the end of each time block had become increasingly more stable as the developers gained experience with the tools and processes and as they factored in feedback from the developers of the individual systems.

JSSEO and the military Services established a Community Share site where each participant can post improvements or changes to the model compiler being used to transform the PIM to the PSM. This approach was intended to foster reuse in the community so that individual programs that are responsible for generating the PSM could leverage the experiences of others.

One of the key difficulties concerns the boundary between the functionality that SIAP provides and the functionality that is the responsibility of the weapon system program office and its contractor. New starts can integrate the JSSEO-provided functionality directly, but for systems that are in development or being upgraded, integration requires the removal of existing functionality and its replacement with new code based on the IABM. Furthermore, the boundary between such joint functionality and the organic functionality is not obvious, and defining it remains a challenge.

Initial IABM releases were intended primarily to familiarize the weapon platform contractors with the IABM tools and processes. As the IABM time box releases progressed, they were expected to culminate in a series of spiral releases, termed "Configurations." The development team tested, verified, and validated each spiral release before it was provided to the Services for incorporation into their particular systems.

The first formal IABM release took place in September 2005 with the release of Configuration 05. Time Box 30 was released in March 2006 and IABM Configuration 07 was initially scheduled for release in September 2007. Funding cuts, however, caused that release to slip into 2008, and it was then being referred to as SIAP Capability Drop 1.

7.4.2 Demonstration and Test

As the Services and their contractors gained experience with the IABM releases, they first integrated the code in their laboratories and then began conducting demonstration and test events. Several of these, including both multi-Service and single-Service events, have taken place since 2006. In December 2006, platform-specific instantiations of IABM were integrated into representations of Army, Navy, and Air Force systems to demonstrate that IABM could operate in a multi-Service, multi-platform environment. In February 2007, the Navy held an at-sea demonstration in which IABM was integrated into the Navy's Common Network Interface. In June 2007, the Marine Corps held a demonstration of IABM integrated into its Tactical Air Operations Module; and in October, the Air Force demonstrated the ability to share the air picture between two Battle Control Systems nodes that had integrated IABM code.

SIAP has also held distributed HWIL test events leveraging the JDEP developed tools[28] (see Section 7.3.2.3). In June 2007, a multi-Service event including the Air Force Airborne Warning and Control System, the Navy Airborne Early Warning and Command and Control Aircraft (E-2), and the Navy AEGIS weapon was the first to test IABM platform-specific implementations in a heterogeneous distributed environment.[29] Another such event was scheduled to take place in late 2009.

7.4.3 Challenges of Weapon System Integration and Synchronization

The scope of the SIAP effort eventually encompassed joint track management, combat identification, and battle management functions to be embedded into targeted weapon systems. Initially, the JSSEO identified ten weapon programs across the Air Force, Navy, and Army, referred to as "pathfinder" programs, which agreed to accept Configuration 05 and became the focus of integration activities. Contractors from these programs were involved in developing the behavioral model as well as in reviewing and providing feedback on the product of each time box. However, considerable concern was raised across the community about the viability of Configuration 05 to satisfy near-term needs. In fact, representatives of several of the military Services indicated that they were exploring alternatives to some of the SIAP capabilities so that they could field near-term solutions. As the expected date of release for Configuration 05 drew closer, it became apparent that only one of these pathfinder programs had actually allocated funding for the integration of the IABM, and even that was to take place in a laboratory environment rather than in a fielded version.

These concerns were not necessarily new, but they reveal an unresolved problem for SIAP. From the beginning of the effort to build an IABM using the MDA approach, stakeholders often raised issues about the interrelationship between the IABM development efforts and their impacts on the target programs. For example, they expressed concerns about the maturity of the MDA development

> SIAP encountered continuing stakeholder concerns about IABM process, products, and impact on performance of the receiving weapon systems

process, the scope of the IABM relative to that of the weapon systems in which they were to be fielded, and the extent to which that software would impact the performance of the receiving systems. Thus, while JSSEO may have succeeded in forging a technical consortium of developers, this did not necessarily translated into a service commitment to implement the results.

7.5 Formalizing the SIAP

Program Management Reviews held in the spring and summer of 2005 highlighted the lack of agreement among the stakeholders about the fundamental capabilities to be developed and the top-level architecture that would drive their development. Consequently, a Joint Architecture Working Group, composed of representatives from the Services, the Missile Defense Agency, and the JSSEO, was established to develop an agreed-to, top-level architecture, focusing specifically on a joint track management capability. This was intended to introduce more top-down structure into the design and development process. Once the stakeholders agreed to this architecture, the next step was to assess the current IABM against it, identifying design gaps and evaluating potential alternatives.

In parallel, the Office of the Secretary of Defense (OSD) initiated an effort to develop a long-term management plan. The intent was to establish management structures and approaches that would more readily lead to Service buy-in. It was felt that such management approaches were necessary to align expectations, synchronize program activities, and enable system engineering across program and service boundaries. They were also needed to allocate resources and coordinate the fielding of the various receiving systems.

In March 2005, SIAP was designated a Special Interest Program; and in March 2006, the JSSEO went before the Defense Acquisition Board, which formally approved continuation of the SIAP program.

During the second quarter of FY 07, the JSSEO formally transitioned to a Joint Program Office under a designated SIAP Joint Program Executive Officer. This second organizational restructuring was intended to formalize the production of necessary acquisition documentation as part of an effort to provide more system engineering rigor into the development of the SIAP capability, to establish funding responsibilities for development and validation of the SIAP joint product, and to support system engineering, risk reduction, and integration of these products into service weapon systems.

While discussions continued with the Services about the specific requirements that Capability Drop 1[30] would include, SIAP efforts focused on developing and delivering this release later in 2008. Of the ten pathfinder programs, few had the necessary funding to be able to integrate the platform-independent modules into their platforms; therefore, the SIAP integration efforts centered on the funded pathfinders.[31]

7.6 Summary

The SIAP approach to system engineering is clearly not the traditional top-down process in which requirements drive specifications and specifications drive development. In SIAP, the requirements were collected at the same time that efforts began to develop capabilities. Recent management direction is to have the SIAP efforts conform more closely to traditional systems engineering and acquisition practice. However, SIAP itself constitutes essentially a horizontal development and integration process, while traditional systems engineering and acquisition processes are vertical and, to a large extent, stove-piped, in that they focus on each system separately.

SIAP has put some fundamental elements in place, including the consortium to develop the IABM and an established process that incrementally develops the IABM in 12-week time boxes. This process is documented in an internal "Deskbook." Target system developers routinely provide feedback to the developers of the IABM. Moreover, early engineering demonstrations and distributed simulations have shown that IABM can, in fact, be integrated with weapon platforms.

While considerable progress has been made, risks and unknowns remain, and SIAP continues to reexamine fundamental processes and driving assumptions. Several significant problems continue to plague the SIAP effort. The integration of the SIAP functionality into multiple target systems is a daunting task, and viable methods have yet to be worked out. Service commitments to implementing SIAP have been generally weak from the outset. While stakeholders have concurred with the fundamental premise of SIAP, they have resisted implementing first the Block recommendations and later the IABM. In some cases, they have questioned the process and the documentation. In other cases, they have failed to commit the level of resources necessary to execute.

Postcript: The SIAP program was officially terminated at the end of FY2009

Finally, the use of MDA as a development approach is still under scrutiny. The MDA process itself is still maturing and its expected benefits have yet to be realized; as noted, only one program funded the integration of Configuration 05 in a laboratory environment, and only a limited number of pathfinder programs have the necessary funding to integrate Capability Drop 1 into their platforms.

The MDA approach represents a significant departure from the initial SIAP program, and it remains to be seen whether it can accomplish its objectives. If the SIAP system engineering effort can, in fact, lead to a shared air picture and do so with enough community-wide buy-in that the individual program managers commit to implementing the common computerized specification, then this effort will have revolutionized the implementation of software distributed across many platforms. Moreover, if SIAP can generate an acquisition approach that accommodates both joint and system-specific functions, processes, and schedules, then it will have made a significant contribution to the development of systems of systems.

This case study demonstrates that while the technical issues are challenging to solve, in many cases, political, organizational "turf," and economic issues are even more difficult than solving the hard technical problems. In fact, we have observed that it is not unusual for "soft" issues to be disguised as hard technical issues.

7.7 SIAP Mapping to the Systems Engineering Profiler

Figure 7.4 shows the Systems Engineering Profiler introduced in Chapter 5 applied to the SIAP effort, with results plotted in the form of a spider chart. The polar chart format allows the reader to highlight those aspects of the SIAP effort that are more

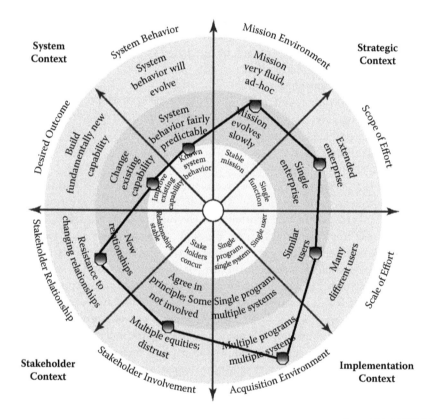

Figure 7.4 SIAP profile. (From Stevens, R., 2008. *Profiling Complex Systems*, Proceedings of the IEEE International Systems Conference, Montreal, Canada, April 2008. With permission.)

akin to traditional systems engineering as well as those that may be less amenable to such processes and techniques.

From the perspective of its *system context* and *strategic context*—the upper hemisphere—SIAP appears to be in line with traditional, well-bounded development efforts. In terms of *desired outcome*, the program does not intend to develop new functionality; track management, for example, already exists today in different weapon systems. Rather, SIAP is designed to fix some well-recognized and long-standing problems by providing a common, computerized specification. The program's fundamental goal is to develop a highly predictable capability that is consistently implemented. Its underlying architecture and fundamental approach are intended to allow the application logic to persist even when the implementation environment changes. From a technical perspective, SIAP seeks to drive out the complexity and unpredictability and move yet closer to the linear system end of the continuum.

From a *system behavior* perspective, SIAP focuses on discrete functionality that lends itself to modeling and simulation. In fact, the program has emphasized a distributed simulation approach to verification and validation. Admittedly, however, what remains to be determined with the necessary degree of confidence is the impact of the IABM-based software on the performance of the target weapon system.

From a *mission environment* perspective, the transformation from independent, geographically based air and missile defense operations to a theater-wide, integrated capability has been agreed to for over a decade now. So has the need for an integrated capability that provides operators with a consistent and complete set of information about airborne entities of interest throughout the engagement zone. While specifics of air and missile engagements will necessarily vary with the particular military operation, these basic operational drivers are not likely to change substantially in the foreseeable future.

It terms of its *scope*, SIAP can be considered to operate in a single enterprise: that of theater air and missile air and missile defense. However, the underlying motivator for SIAP was the discrepancies among systems developed and operated by different Services, which resulted in a confused air picture and, at worst, could lead to incidents of "fratricide." To the extent that the Services continue to view themselves as separate enterprises (a situation unlikely to change soon), one can make the case that SIAP is at the border between a single enterprise and an extended enterprise.

But when viewed from the perspective of its stakeholders and the acquisition context—the bottom hemisphere of the Systems Engineering Profiler—SIAP is more akin to the mega-systems discussed previously.

From the perspective of *scale*, while there may be as many different types of users as there are specific systems that participate in this enterprise, the functions supported by SIAP are essentially the same, independent of platform.

From the *acquisition* perspective, successful execution of the SIAP capability clearly depends on its integration into multiple systems, each of them separately managed, funded, engineered, and acquired. While SIAP has identified the affected weapon systems and designated "pathfinder" programs in each of the Services, actual integration has yet to occur. A long time may elapse before it is possible to determine the impact of these separate integration efforts on the ability to generate the needed single integrated air picture.

The acquisition context is clearly linked to the stakeholder context. Because of the particular nature of this program, there are several critical stakeholders over which the SIAP program has had little direct control, although they do have the ability to influence. SIAP has established a technical consortium that, by all accounts, is effective in developing the IABM and has funded participation by these stakeholders. Yet tensions remain. While the most recent changes in the SIAP management structure will formalize relationships among the stakeholders, the extent to which the SIAP program will succeed in forging the collaborative acquisition environment critical for implementing the SIAP vision remains unknown.

7.8 Insights for Engineering Mega-Systems

- *Requirements.* It is difficult to define the requirements right the first time—or even perhaps the second or third time. Even when requirements are considered "right," they will change as new concepts of operations evolve.
- *Specifications.* Current paper standards and specifications do not guarantee common implementations. The more complex the standard or specification, the greater the likelihood of variances in interpretation and implementation.
- *Analysis of systems of systems.* Federations of simulations offer an attractive approach to analysis of systems of systems because of their ability to explore interactions among authoritative simulations of systems already developed by different organizations.
- *System engineering process.* A robust and adaptive system engineering process is needed to ensure that what is developed responds to the expectations of the users, even as those expectations evolve over time. The architecture and development path should be designed to accommodate evolution of functionality.
- *Stakeholder concurrence.* The greater the diversity of programs involved, the greater the need—to achieve stakeholder concurrence and architecture convergence—and the greater the difficulty of achieving it. If those who will be responsible for implementation are involved in development, they will be more likely to commit to using the system. However, technical consensus is necessary but not sufficient. There must also be buy-in from the receiving programs and their proponents and users.
- *Balance between top-down and bottom-up.* Systems engineers must balance the extent to which they guide and enforce from a top-down perspective and how much they encourage ideas to "bubble up" from the bottom.

- *Strategy and implementation.* Not only must formal implementation follow strategy, but strategy must also be informed by early implementation.

Endnotes

1. The information for this case was based on program documentation and interviews with the following people, who were generous with their knowledge and insights: Dr. Michael Bienvenu, Randy Burnham, Sandra Cole, Ellen Conway, Kimberly Crider, Dr. Judith Dahmann, C. Zachary Furness, John Nordmann, Thomas Nyman, and Karen Rigopoulos.
2. In particular, Network Participation Group #7 (the Joint Surveillance Picture).
3. The international standard for Link-16 is NATO STANAG 5516. The U.S. near-equivalent is MIL-STD 6016B. The United States is a signatory to the STANAG, which carries the force of a treaty. The two reference standards are not completely identical. The MIL-STD contains a number of U.S.-only standards. In addition, the MIL-STD is constantly being amended and refined. The STANAG follows after the United States has adopted modifications.
4. The term "joint" connotes activities, operations, organizations, etc. in which elements of two or more Military Departments participate. http://www.dtic.mil/doctrine/jel/doddict/data/j/02803.html.
5. A commander of one of the unified or specified combatant commands established by the President. http://www.dtic.mil/doctrine/jel/doddict/data/c/01044.html.
6. The Joint Data Network (JDN) is the collection of near-real-time communications and information systems used primarily at the coordination and execution levels. The backbone of the JDN is Link 16 (a message standard) transmitted via Joint Tactical Information Distribution System (JTIDS) and Multi-function Information Distribution System (MIDS) terminals.
7. SIAP uses the term "system engineering." As such, this term is used in this book when referring to SIAP engineering efforts. In other portions of the book, the term "systems engineering" is used.
8. The JROC, as it is known, is an advisory council to the Chairman of the Joint Chiefs of Staff established to identify and assess the priority of joint military capabilities. It consists of the Vice Chairman of the Joint Chiefs of Staff, who serves as the chairman; the Vice Chiefs of Staff of the Army, Navy, and Air Force; and the Assistant Commandant of the Marine Corps. The JROC is responsible for reviewing, approving, and providing guidance on all requirements matters that have a joint flavor and exceed a certain funding level. This includes setting budgeting priorities and amounts for operational capabilities.
9. *The* C4ISR *Department of Defense Architecture Framework* provided initial guidance on architecture products. It identifies three sets of products that correspond to operational, system, and technical views. This document was superseded by the *DoD Architectural Framework,* which applies to *all* DoD systems, not just command, control, communication, intelligence, surveillance, and reconnaissance systems. The current version of the DoD Architectural Framework is Version 2.0, dated May 2009.

10. "As-is" architectures document existing components, processes, and the interactions among them.

11. For example, at a meeting of the SIAP Oversight Council held in early spring 2001, the principals indicated a desire to be able to demonstrate some "near-term" impacts for the JROC prior to the scheduled 1 December recommendations briefing. They urged the SIAP System Engineer to look for additional opportunities to make early inroads into current problems.

12. In addition, the JFCOM memo identified several other areas of concern and asked the Task Force to provide a technical assessment with recommendations within 90 days. The issues identified included a method to provide feedback to the combatant commanders about numerous Link "link 16" deficiencies identified during operations so that the commanders can ensure that "bugs" are being fixed. The JFCOM also suggested an integrated database describing data link function implementation, and certification across all members of the TAMD family of systems was also noted. Finally, the JFCOM expressed concern that it takes too long to fix problems. They asked that the SIAP SETF include in its Block 1 recommendations, options for timely implementation of any proposed improvements.

13. Examples of types of tasks include validating and populating track accuracy databases, conducting functional decomposition, developing behavior models, and assisting in planning and executing throughput analysis, among others.

14. Modeling and simulation tools used for analysis were the Air Defense Simulation (ADSIM), Ballistic Missile Defense (BMD) Benchmark, Missile Defense Wargame Analysis Resource (MDWAR), Track Accuracy Parametric Model, and the Operational Data Driven Simulation for Correlation Algorithm Performance Evaluation (ODDSCAPE).

15. The SIAP SETF conducted Block 1 HWIL analysis in the Joint Distributed Engineering Plant (JDEP) in a limited scenario as a proof of process.

16. Empirical analyses were conducted on data collected from the All-Service Combat Identification Evaluation (ASCIET) 00, Joint Combat Identification Evaluation Test (JCIET) 02—new name for ASCIET, and Roving Sands 01 live events.

17. MDA is a specification of the Open Management Group (OMG), an open-membership, not-for-profit consortium that produces and maintains specifications for interoperable enterprise applications.

18. In the field of software engineering, the Unified Modeling Language is a standardized specification language for object modeling. UML is a general-purpose modeling language that includes graphical notation used to create an abstract model of a system, referred to as a UML model.

19. Just as the PIM is a model, the means by which the PIM is transformed into the PSM is also a model.

20. Examples of ensemble behavior include error handling and data consistency.

21. In the near term, the JDEP was to identify and fault isolate interoperability problems of fielded or soon-to-be-fielded JTAMD systems, isolate faults, and test fixes to these systems.

22. The Navy Distributed Engineering Plant (DEP) used data networks to connect various Navy designs, testing, and training facilities to create a distributed engineering environment that enabled actual hardware and software to be integrated at the laboratory level. The Navy DEP was used to isolate faults and verify resolution of interoperability Battle Group interoperability problems.

23. The JDEP technical framework allows a set of distributed components, based on a set of basic industry standards, to be configured into federations and tailored to the needs of the particular user.
24. The first event was a pilot both for the common SIAP tools and the JDEP framework. In the event, set in a hardware-in-the-loop environment, the actual mission computer was driven by a set of inputs simulating radar, sensor, and tactical data messages.
25. This is intended to validate that the simulation environment has the requisite fidelity.
26. Time boxes are now delivered approximately every 12 weeks.
27. The tool supports an Action Semantic Language (ASL), which, when combined with the UML model, is sufficient to generate source code (in this case, C++) automatically. In late 2007, SIAP migrated from the Kennedy Carter tool to Rhapsody, citing performance, maintainability, and productivity as the reasons for doing so.
28. Some JDEP tools and infrastructure are now being managed by the Joint Mission Environment Test Capability program, a follow-on program to JDEP.
29. Participants in this distributed test event were in different locations, including Fort Huachuca, Arizona; Pax River, Maryland; Dahlgren, Virginia; Dam Neck, Virginia; and Camp Pendleton, California.
30. Capability Drop 1 is to provide joint track management capability. Battle management capability is to be provided in a subsequent release.
31. Postscript: The SIAP program was officially terminated at the end of Fiscal Year 2009.

Chapter 8

Developing the Electronic Product Code Network

In June 2003, Wal-Mart Stores Inc. announced that it expected its top 100 suppliers to deploy new radio frequency identification (RFID) tags by January 2005. These tags, intended to replace optically scanned bar codes, would be used to track pallets and cases of goods in Wal-Mart's supply chain and thereby further improve inventory visibility. Improved visibility, in turn, was expected to yield greater supply chain efficiencies—smaller inventories, fewer out-of-stock items, and less waste and loss—and thus reduced costs and increased profits.

This case study focuses on the development of the EPC network as a collaborative effort between a number of universities and their industry sponsors. It highlights the path from a top-level vision to demonstrations of technology and development of business cases. It also illustrates the consequences of underestimating critical stakeholder concerns, in this case concerns by a small but quite vocal group about privacy issues.

While the initial technologies have been developed and implementations have begun, it remains to be seen whether the technologies and the community approach to developing them achieve the expected magnitude of change across the global supply chain.

The particular RFID system called for by Wal-Mart was developed by the Auto-ID Center, a Massachusetts Institute of Technology (MIT)-led, industry-funded, global consortium of universities established to conduct RFID research. Over a period of 4 years, this academic research project developed and specified the technologies and demonstrated them in a series of large-scale field trials. In early November 2003, the effort transitioned from a university-led research activity to an industry-funded effort to develop stable standards and drive toward global, multi-industry adoption of these technologies. At the same time, the various Auto-ID Centers were renamed Auto-ID Labs.

This case study focuses on the development of the Electronic Product Code (EPC) Network, an open, global system to track individual items using low-cost RFID tags. Not only is this technology potentially transformational for the supply chain, but the organizational construct is itself also revolutionary. As stated in one volume of the Auto-ID Center's *Technology Guide* (Auto-ID Center, 2002), "the Auto-ID Center may be the first time in history that companies from different industries and different regions of the world have come together to develop technology they feel would benefit their businesses—and their competitors' business."

8.1 Background

"Automatic identification" is the umbrella term given to technologies used to help machines identify objects. It encompasses a wide range of technologies, including bar codes, smart cards, voice recognition, optical character recognition, and RFID tags.

Perhaps the most common of these technologies is the bar code. While the initial research began in the late 1940s and the first patent was granted in 1952, it was not until 1969 that the first two systems were installed: one in a General Motors plant to monitor axle production and the other in a distribution facility in New Jersey to direct shipments to the correct loading bays. Around the same time, two branches of the food distribution industry—suppliers and grocers—held a series of meetings that culminated in a decision to seek a standard "inter-industry" code. In 1974, after a 4-year effort, the Universal Product Code was adopted as a standard (GS1 US, 2006).

While bar codes are clearly not going away, they have some well-recognized limitations (Lulay, 2003). Specifically, they require line of sight for scanning, and each bar code must be read individually. Additionally, the label itself has constraints: It can store only a very limited amount of information and it lacks a read/write capability. Finally, because the label cannot be read automatically, but requires a person to scan it, the process is subject to human error.

As the 25th anniversary of the bar code approached, the two organizations responsible for administering the bar code standards—the Uniform Code Council (UCC) in the United States and the European Article Numbering Association (EAN)[1]—were looking for next-generation identification technologies. In 1998, they established a project team to explore RFID technologies,[2] document business cases, define technical requirements, and map these requirements to the potential technologies. A White Paper published in November 1999 (EAN UCC, 1999) concluded that while RFID technologies presented opportunities for application to supply chain management, the lack of open standards posed a critical barrier. RFID technologies, such as those used in the Exxon Mobil Speedpass®, were notoriously proprietary. Furthermore, cost, while not a significant factor for then-existing applications, would be a critical factor if RFID tags were to be used as ubiquitously as bar codes.

At the same time that the UCC and EAN were beginning to explore RFID tags to complement existing bar codes, two other factors were converging: a business need and a technology concept. The business need came from the supply chain of the consumer products industry. Two major producers of consumer package goods, Proctor & Gamble and Gillette, were experiencing problems that were costing them significant revenue.

For Proctor & Gamble, the key issue was to optimize product availability by reducing "out-of-stock" levels, which represented $3 billion a year in lost revenues for the corporation (Kirsner, 2002). A 2003 report estimated that worldwide, retailers could be losing up to 4% of sales due to out-of-stock merchandise. The causes span the entire supply chain. Products were ordered but did not get to the store. If ordered and received, they did not reach the shelf. At the shelf, they were stolen or not replenished rapidly enough. Ironically, at the time, 33% of the out-of-stock items were actually available at the retailer's distribution center (or warehouse) (Lulay, 2003).

For Gillette, the issue was to minimize inventory losses—shrinkage[3]—particularly associated with Mach 3 razor blades. Because they are small, in demand, and of high value, these blades (along with batteries, compact discs, and digital video disks) are among the items most often stolen. Gillette estimated that Mach 3 razors have a 5% shrinkage rate, amounting to $180 million per year in lost revenue (Lulay, 2003). Thus, both Proctor & Gamble and Gillette were looking at innovative solutions that could help them track items in the supply chain.

The second factor was a technology concept that emerged from the intersection of two technology trends. The first trend was the explosive growth of the Internet and, with it, of networking and embedded control. In the mid-to-late 1990s, enterprising engineers were networking all manner of household devices, including coffee pots and microwave ovens. The second trend was the emergence of new tagging and sensing technologies; for example, Exxon Mobil's Speedpass system was introduced in 1997.

In 1998, two researchers at MIT, Dr. David Brock and Professor Sanjay Sarma, and their colleagues were dealing with the challenge of getting a robot to "see" and identify objects around it—a complex task requiring the robot, in effect, to mimic human sight. Brock reversed the problem and proposed that instead of focusing on improving the robot's ability to sense its environment, research should center on enabling the objects in the environment to identify themselves to the robot. Each object would have a unique identifier that would not only identify it, but also describe its characteristics so that the robot would "know" how to interact with it. The researchers at MIT began exploring the potential of low-cost RFID tags for this application because, unlike traditional tagging technologies such as the bar code, RFID tags did not require line of sight but could be read as long as they were within the range of the reader. Also unlike bar codes, RFID tags made it possible to identify individual objects, not just classes of objects.

MIT researchers proposed moving functionality from the chip to the network; leveraging and building on accepted Internet standards; and, following the Internet model, ensuring that these standards were open and non-proprietary.

That idea triggered the development of a system-level approach to automatic identification centered on networking individual items, each equipped with a low-cost RFID tag that would contain a unique identifier. For these tags to be used at the individual item level, they would have to cost on the order of pennies. However, such a tag could have only limited on-board functionality if it were to remain so inexpensive. The MIT researchers proposed moving functionality from the chip to the network, leveraging and building on accepted Internet standards, and, following the Internet model, ensuring that these standards were open and non-proprietary.

8.2 The Auto-ID Center

The convergence of these three separate interests led to the establishment of the Auto-ID Center at MIT on 1 October 1999. The Auto-ID Center, located in the Department of Mechanical Engineering, had the mission to develop and promote open Auto-ID standards. Initial funding came from MIT, Proctor & Gamble, Gillette, and the UCC, along with the EAN, its international counterpart. Kevin Ashton took a leave of absence from his position as a product manager at Proctor & Gamble and came to MIT to serve as the Executive Director of the Center— probably the first time that someone outside of academia has held such a position. Professor Sunny Siu of MIT became the first Research Director and Alan Haberman of the UCC became the Chairman of the Board of Overseers.

RFID was the product of a global consortium of industry and academia.

The Auto-ID Center quickly grew into a global consortium of industry and academia headquartered at MIT. It included

- Research institutions at universities around the world
- An Auto-ID Board of Overseers composed of the end-user sponsors and standards bodies
- An Auto-ID Technology Board composed of vendor sponsors

Fees from end-user and vendor sponsors funded the research. Standards bodies paid no fees.

While MIT remained one of the key research institutions and continued to serve as the headquarters for the consortium, other university-based Auto-ID Centers were added. In 2000, Cambridge University in the United Kingdom joined. Subsequently, Auto-ID Centers were established at Adelaide University in Australia; the M-Lab at St. Gallen, ETH Zurich, in Switzerland; Keio University in Japan; and Fudan University in the People's Republic of China. Different centers concentrated on particular research problems, with MIT focusing on the infrastructure, initial applications, and initial business cases, and Cambridge, for example, focusing on manufacturing and systems control applications as well as issues related to the extended infrastructure.

Industry sponsors represented the potential end users of this system and encompassed a wide variety of Fortune 500 product manufacturing and retail organizations, along with government organizations. Starting with just three sponsors and focusing exclusively on the consumer products industry, the list of industry sponsors (see Table 8.1) grew rapidly and expanded to encompass major retailers and representatives of other industries, such as pharmaceuticals and paper manufacturing. It is worthwhile noting here that the Auto-ID Board of Overseers represents an obvious case of "unlikely partners." Many of these sponsors are direct competitors, yet they were willing to collaboratively sponsor and fund an effort that would not give them an individual competitive edge. They believed that achieving the vision of RFID across the global supply chain required a set of open standards that would be available to all.

Vendors of RFID-related products and services also joined as sponsors and were represented on the Auto-ID Technology Board (see Table 8.2). These included not only well-established technology companies, but also start-ups seeking business opportunities in what they saw as a growing and lucrative new business market.

8.3 The Vision: An Internet of Things

The vision of the Auto-ID Center was audacious: no less than to revolutionize how products are made, bought, and sold by merging the world of atoms (things and people) and the world of bits (information). Moreover, the designers wanted to do it globally. In their vision, physical objects would communicate with one another in real-time, and trading partners would know exactly where their products were at every point in the supply chain. This knowledge, in turn, would allow fundamental changes in how the supply chain operates, promising increased efficiencies not only to business, but also to consumers.

> "We're looking at a world in which computers will know about things without having to be told by human beings."
>
> **—Kevin Ashton**
> *Auto-ID Center Executive Director*

Table 8.1 Auto-ID Center Industry Sponsors

Abbott Laboratories	International Paper	Sara Lee Corporation
Ahold IS	Johnson & Johnson	Smurfit-Stone Container Corp.
Best Buy Corporation	Kellogg's Corporation	Target Corporation
Canon, Inc.	Kimberley Clark Corp.	Tesco Stores Ltd.
Carrefour	Kraft Foods Inc.	Toppan Printing
Chep International	Lowe's Companies, Inc.	Uniform Code Council
Coca-Cola	Mead Westvaco	Unilever
CVS	METRO AG	United Parcel Service
Dai Nippon Printing Co., Ltd.	Mitsui & Co., Ltd.	Unites States Postal Service
Department of Defense	Nestle	Visy Industries
EAN International	Pepsi	Wal-Mart Stores, Inc.
Eastman Kodak	Pfizer	Wegman Food Markets, Inc.
Gillette Company	Philip Morris International (Philip Morris USA)	MeadWestvaco
Home Depot	Proctor & Gamble Co.	Yuen Foong Yu Paper Mfg.

Early on, the MIT researchers imagined a future world in which a unique identifier would accompany each product from its manufacture to the point of sale and finally to its disposal. Tags placed on individual items, cases, and pallets would be scanned automatically by automated readers located throughout the supply chain, thus allowing these items to be identified, counted, and tracked without human intervention. Once the items reached the store, readers integrated into shelves would report stock levels and facilitate automated replenishment, and other readers mounted at store doors would automatically identify all items in a shopping basket, thereby enabling customers to bypass the cashier. For the home, the researchers envisioned systems such as smart refrigerators that would be able to track their contents, alerting the homeowner when items were consumed or when they were approaching their expiration date. At recycling centers, the tags would allow the items to be automatically sorted and possibly rerouted to their manufacturer for reuse.

Table 8.2   Auto-ID Technology Board

Accenture	Information Resources, Inc.	RF Saw Components
ACNielsen	Intel	SAMSYS
Alien Technology	Intermec	SAP
Avery Dennison	Invensys PLC	Savi Technology
AWD	Ishida Co., Ltd.	Sensitech
British Telecommunications (BT)	KSW Microtec AG	Sensormatic Electronics Corp.
Cash's	Manhattan Associates	Siemens Dematic Corp.
Catalina Marketing Corp.	Markem Corp.	STMicroelectronics
Checkpoint Systems, Inc.	Matrics	Sun Microsystems
ConnecTerra, Inc.	Morningside Technologies	TAGSYS
Flexchip AG	NCR Corporation	ThingMagic
Flint Ink	Nihon Unisys Ltd.	Toppan Forms
GEA Consulting	Nippon Telegraph and Telephone Corporation	Toray International, Inc.
GlobeRanger	NTT Comware	Vizional Technologies
IBM Business Consulting Services	OATSystems	Zebra Technologies Corporation
IDTechEx	Philips Semiconductors	
Impinj, Inc.	Rafsec	

A Gartner Strategic Analysis Report (Magrassi and Berg, 2002) outlined the possible benefits of such intelligent products:

- Automated proof of delivery
- Self-managed products for transportation (i.e., smart packages that "know" where to go)
- Self-picking products that "know" their own due dates
- Automated "on-shelf" availability
- "Use by" information, end-of-life management, and recycling

To become reality, this vision had to be predicated on an Internet and World Wide Web-like approach in which a limited set of well-defined and commonly accepted standards fuel information sharing by strategic business partners on a global level.

8.4 Concept and Technologies

The MIT concept revolved around the ability to identify all discrete items—every can of soda and every roll of paper towels—and consisted of elements centered on the development of very low-cost RFID chips. It included

- A numbering scheme capable of identifying all global items uniquely
- A pointer to data about the item on a network
- A strategy for leveraging the network to carry and store vast amounts of data
- The "bet" that the resulting smaller chips could be manufactured at substantially lower costs[4]

The concept comprised a data capture portion and an intelligent infrastructure. The data capture portion encompassed the numbering scheme, referred to as the Electronic Product Code (EPC), low-cost tags, and low-cost, networked readers. The intelligent infrastructure included software to manage and filter the data captured from tags (initially called Savant and subsequently referred to as the Application-Level Events software standard), the EPC Information Service, the Object Name Service, and the Physical Markup Language. Figure 8.1 provides an overview of the architecture components and their relationships.

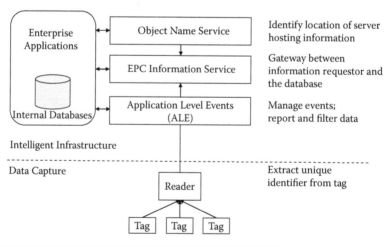

Figure 8.1 EPC network architecture.

8.4.1 Data Capture Portion

The EPC is a universal standard numbering scheme, the equivalent of a license plate for each item (see Figure 8.2). Like the bar code, the EPC is divided into numbers that identify the manufacturer and the product. However, the EPC adds an extra set of numbers that is akin to a serial number and is used to identify unique items (such as the individual cans of soda mentioned earlier). In addition to being a unique identifier, the EPC serves as an addressing scheme for product data locations, similar to the use of Internet Protocol (IP) addresses.

The EPC header distinguishes between types and versions of the EPC. The initial specification developed by the Auto-ID Center called for an interim 64-bit code as well as a 96-bit code. The smaller interim code was designed to provide enough unique identifiers for early applications and to help keep down the cost of the RFID microchips. In 2003, the Auto-ID Center proposed a 256-bit format designed to be used as a universal identification scheme rather than being limited to use in identifying physical items.

The code is carried on the RFID tag, which consists of a microchip attached to an antenna. To drive down the cost of the tag to a target value of 5 cents (see Section 8.3), the Auto-ID Center focused on passive tags. Unlike active tags, which have a battery, passive tags draw power from the reader, thus making them cheaper to produce and requiring no maintenance. The drawback is that passive tags have a shorter read range—less than 10 feet, versus the 100 feet or more of active tags. Technology sponsors emphasized research to devise ways to manufacture smaller, lower-cost chips and innovative antennas.

In the EPC scheme, low-cost readers send out electromagnetic waves that power the RFID tag, enabling it to transmit the information on the chip back to the reader. The Auto-ID Center focused on creating a network of readers and developed the design for an agile reader (one able to work at different frequencies) that can read tags from a distance of around 4 feet.

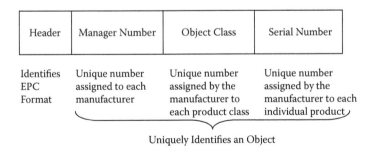

Figure 8.2   Structure of the Electronic Product Code (EPC).

8.4.2 Intelligent Infrastructure

The second portion of the network architecture is the intelligent infrastructure that leverages and builds on the Internet architecture. This network provides a standard-ized way of exchanging EPC data, both within a company and between strategic partners in the supply chain. The Application-Level Events software is a technology designed to manage and filter the flow of potentially vast quantities of data and thus avoid overloading the network. The filters are organized hierarchically. Those associated with readers are responsible for filtering and logging in data, adjusting for incorrect or missed readings. At higher levels, filters compare readings from separate readers and adjust for duplicates. At each level, the filters interact with one another, passing along information: For example, a filter at a distribution center could inform a store filter that a shipment has been dispatched.

The EPC Information Service (EPCIS) serves as a gateway between requestors of information and both internal and external databases. EPCIS is a specification for a standard interface that allows trading partners to share and exchange infor-mation independent of the specific databases in which that information is stored, their underlying operating system, programming language, or a vendor's particular legacy information system.

The Object Name Service provides a pointer from the EPC to a location on the network where information about that product is stored. It is analogous to the Domain Name Service used to point computers to sites on the World Wide Web. The databases serve as the information archives that store data records about physi-cal objects and are linked to the identification number.

EPC uses the Physical Markup Language (PML) (Brock, 2001; Floerkemeier et al., 2003) as a common standard for describing physical objects, and their charac-teristics and state. It includes static information, such as dosage, shipping, expiration, and recycling information. In addition, its intended use is to provide instructions for machines, such as microwave ovens and laundry appliances in homes or machine tools and industrial equipment in factories, about how to handle the product. The original vision was also designed to include dynamic information that might describe how an object behaves under different circumstances. The PML uses the eXtensible Markup Language (XML) as the method of storing and transmitting data.

8.5 RFID Design Process and Implementation

The researchers at MIT specified and developed most of the key elements of the EPC system very early in the project. In fact, MIT succeeded in building proto-types within weeks of the original concept to demonstrate that the underlying ideas were viable.

Initially, the MIT team did not limit its vision of the EPC system to any particular business area or problem space. It was only after they started working with Kevin Ashton that the group began to focus the broad vision onto the specific needs of one particular industry: the consumer packaged goods industry. From 1999 through 2001, research teams concentrated on designing the prototype system software and hardware. Starting in 2001, they initiated a series of field trials to test the technologies and demonstrate their value.

8.5.1 Architecture and Design

While the EPC had no formal set of requirements, such as those articulated for traditional government acquisition projects, it did have a de-facto design target. Specifically, one of the senior managers at Proctor & Gamble told con-

> The "5-cent tag" served as the de-facto design target.

sortium leaders that if they could produce a 5-cent tag, he would put it on every roll of Bounty paper towels. The 5-cent tag thus became the driving requirement for the design of the entire system. To produce such a low-cost tag, researchers focused on two complementary approaches: developing new, low-cost manufacturing techniques while at the same time limiting the amount of data that could be stored on the tag itself. What could not be put on the tag had to be put on the network. This led to the emphasis on the data structure and network aspects of the Auto-ID system.

Just as there was no formal requirements document, the EPC also had no formal architecture products[5]—at least not the kinds of architecture products called for in federal acquisitions. However, the program certainly had an envisioned concept of operations and even an operational architecture as well as a system architecture that evolved over time (see Figure 8.1). Although this operational architecture was not labeled as such and was not developed at the outset to drive the design, it graphically described how the Auto-ID technologies envisioned by the consortium would be used to automate the supply chain. The Auto-ID Center actually created that graphical description as a marketing tool to communicate the concept to audiences outside the consortium. In the parlance of the architecture community, it was, in effect, a "to-be" operational architecture. There was also a technical architecture, again not documented as such, heavily based on adaptation of Internet standards.

> What turned out to be complex was not the technologies themselves, but rather their interactions with the social and business processes.

The technologies used were individually rather simple. For example, the MIT researchers developed the basic data structure of the EPC in a single afternoon, and that data structure has essentially remained unchanged to date (Brock, 2003). Nor were the networking aspects of the Auto-ID systems especially complicated. In fact, the researchers mimicked and adapted some well-known Internet protocols. Although the Auto-ID, unlike the Internet, was to scale up to identify and track every item globally the researchers did not consider scaling a difficult problem. They saw several available options to deal with the issue, both in terms of how the data would be structured and how it could be managed as it moved through the system. In fact, what turned out to be complex were not the Auto-ID technologies individually, but rather their *interactions with the social process and the business process* (see Section 8.3.3).

8.5.2 Business Case Research

Parallel to the development of the technologies themselves, the Auto-ID Center established a Business Case Action Group to build a series of business cases that would drive adoption of the Auto-ID technologies. Industry-specific cases examined the consumer package goods industry as well as the retail supply chain, freight transportation, manufacturing, and store operations. For each case, the group sought to define and quantify implementation costs and anticipated benefits, including those that would accrue from reduced costs as well as from increased sales. In all cases, they tried to assess the return on investments and the expected impact on the "bottom line."

The group also chartered the development of several business cases oriented toward market development. They included surveys of potential adopters to gain feedback on planned investments and application priorities, as well as drivers and barriers to adoption. Market sizing analyses and assessments of opportunities in new markets, such as the automotive value chain and even the mass transit market, were also commissioned. In addition, the group developed a series of online tools such as the Auto-ID Calculator, which allowed a potential adopter to model a specific business case and determine whether applying EPC technologies would deliver a return on investment. The Auto-ID Center first published the resulting reports on the sponsor-only portion of the Auto-ID site and eventually made them available to the general public.

> Business cases specific to particular industries were used to drive adoption of the technology.

While business cases are often part of a traditional systems engineering process, they are typically constructed during the initiation phase of a project and are used to determine whether the prospects offer a sufficient basis upon which to proceed. In the commercial world, that basis is often articulated in terms of monetary return on investment. The difference in this situation was that these business cases were developed in parallel with development of the technologies and reflected the critical role played by individual corporations in determining whether, where, and at what pace they would invest in these technologies.

8.5.3 Early Field Tests

In March 2001, the Auto-ID Center assembled a team to plan and implement a series of field tests. These field tests had two objectives: (1) to take the technology from the laboratory to the real-world environment, and (2) to prove the "power and effectiveness of the EPC and to blaze a trail for future adoption" (Albano, Undated). More specifically, the field trials were to demonstrate the system's ability to locate any pallet, case, or item anywhere and at any time in the supply chain, beginning with manufacture and continuing through final disposal or recycling.

Field trials in the United States were structured in three incremental phases, starting with the pallet level in Phase I, adding case-level tracking in Phase II, and then incorporating item-level tracking in Phase III. The field trials were held in facilities operated by members of the consortium, with additional field trials planned for Europe and Japan.

> Early field trials were used to generate interest and gain insights about both technical implementation and real-world operational drivers and constraints.

Phase I began in October 2001, just 2 years after the consortium was formally constituted, and continued through the first quarter of 2002. This phase used existing hardware and focused on assessing the technical feasibility of the Auto-ID-developed software. Phase I started by tracking pallets leaving the Proctor & Gamble factory in Cape Girardeau, Missouri, and arriving at Sam's Club in Tulsa, Oklahoma. It eventually grew to encompass five facilities: one factory, distribution centers belonging to two different manufacturers, one retailer depot, and one retail store (including both the staging area and the retail floor). This first field trial proved to the developers that the software worked as planned and helped them not only learn some lessons about how to implement the technology and how easily it scales, but also gain insights about the actual implementations.[6] Data collected was monitored by the Auto-ID Center at MIT, and participating field test sponsors were invited to view it by logging on to a website.

Phase II began in February 2002 and, like Phase I, used existing hardware. This trial phase was considerably more ambitious in that it not only set out to track cases as well as pallets, but also increased the number of participating sponsors, sites, and products involved. Phase II encompassed a total of two manufacturers, five manufacturer distribution centers, two retail depots, and three separate stores that spanned four separate supply chains. This phase was designed to increase the load on the software network, cope with more realistic problems, test multiple technologies and their interactions, and continue to improve data displays. In addition, the test was intended to generate information for business case development. Like the first field test, the second allowed developers to better understand how to implement the system and provided them with valuable "lessons learned" about the nuances of real operations. For example, they discovered that various factors—including physical size, packaging materials, product composition (liquid, metal, powder, etc.), and location on the package—affected the readability of the tag. They also learned about the impact of different types of installations, interference with existing RF systems, and the need for feedback.

Auto-ID-compliant technologies, including cheap tags and low-cost readers, were first introduced in Phase III. Unlike the first two field tests, which had used surrogate technologies, this phase used tags and readers that met the proposed specifications. All new hardware was first tested to ensure compliance before it was implemented in the field.

Phase III further expanded the scope of the effort, this time introducing item-level tracking and bringing the technology from the back into the front of the store. The focus of the item-level tracker was on applications related to consumer availability, theft prediction and apprehension of thieves, product freshness and out-of-date monitoring, as well as accuracy in end-item stock and replenishment.

As envisioned, the test would not only have involved EPC technologies, but also incorporated end-user initiatives such as the Gillette-developed "smart shelf." This smart shelf was designed to use RFID technologies to scan its contents and, via computer, alert store employees when supplies were running low or when theft was detected.[7] However, plans for a test of the smart shelf, which was to start in June 2003 in a Wal-Mart store in Brockton, Massachusetts, were unexpectedly canceled soon after privacy advocates raised concerns and threatened a boycott of Gillette products. A Wal-Mart representative reported that Wal-Mart had ceased in-store RFID testing, citing the company's intention of focusing on installing the technology in warehouses and distribution centers (C/NET News, 2003). However, similar in-store trials were conducted at retail stores in the United Kingdom and in Germany.

8.5.4 Privacy Pushback

Privacy quickly became a key issue for RFID implementation. Protests by privacy advocates succeeded in terminating several additional front-of-the-store initiatives and are thought to be responsible for refocusing RFID implementations to back-of-the-store uses and delaying the projected widespread use of item-level tracking by about 10 years. In addition, several states and federal agencies held hearings on policy regarding RFID-related privacy issues. At least one state passed legislation regulating the use of RFID technologies, and several others were at one time considering doing so.

> Consumers and privacy advocates were stakeholders with a non-financial interest in RFID developments.

By all accounts, the Gillette smart-shelf trial was only one of several tests terminated because of negative publicity generated by privacy advocates. A proposal by the Italian clothing manufacturer Benetton to embed RFID in labels to track merchandise from the point of manufacture to the point of sale generated threats of a worldwide boycott. These tags were originally intended to remain active after sale so that the store could use them to track returns. Shortly thereafter, Benetton announced that it had changed its plans to put RFID tags in its clothing, and that the company intended to study the technology, including its privacy implications, before taking further action.

Worries about privacy did not result only from in-store monitoring of consumer behavior. Privacy advocates have been particularly concerned about the potential of abuse after the point of sale. They feared that the tags attached to or embedded in the products that consumers buy and carry could be read by massive networks of RFID sensors. They also worried that corporations, individuals, or even governments could use the resulting information not only to monitor what people buy, but also to track their behavior and movements, and that such information could potentially be obtained and misused by hackers and criminals.

From the onset, the Auto-ID Center recognized that privacy would be a key issue and, as early as 2001, conducted an Internet-based survey within the United States. The survey found that 55% of the respondents were either very or somewhat concerned about privacy (Ashton, 2003). In 2002, the Center followed up with a survey of public opinion in the United States, Europe, and Asia to anticipate how the public would perceive the new technology, alleviate any concerns, and explore ways in which the network could be improved (Duce, 2003).

The Auto-ID Center found that consumers generally saw the benefits of the technology accruing exclusively to industry—not to them—and that they worried primarily about the potential for abuse rather than what was actually being developed. More specifically, consumers reported concerns about being tracked, particularly via the clothing that they wear, having companies or the government know what they buy, and forfeiting personal security (e.g., if muggers could know the contents of their shopping bags). They also worried about the potential impact on their health and on labor and unemployment. Allaying these fears was considered very difficult because "they are based on an 'unknown future,' were purely emotional and appeared to be quite deeply rooted" (Duce, 2003). Thus, the Auto-ID Center found that EPC had to confront unexpected consumer resistance stemming from deep concerns about potential invasions of privacy and cynicism about the commitment of both government and business to protecting consumers' privacy and personal information.

The Auto-ID Center's response was multi-faceted, including research initiatives, an external communications campaign, and privacy guidelines. Researchers at MIT and other global Auto-ID Centers began to explore candidate designs that would improve the privacy protections incorporated in the EPC network and identified potential technical research directions (Sarma et al., 2002). Concurrently, the Center brought in a public relations firm that proposed the development of an external communications plan intended, at least in part, to counter opposition to widespread RFID deployment and use. CASPIAN (Consumers Against Supermarket Privacy Invasion and Numbering), a privacy advocacy group, located the proposed plan on the restricted website and subsequently republished it, heightening concerns that the policy was merely a public relations tactic. Although the Auto-ID Center deleted the communications plan from its member-only website, several sites had mirrored this and other Auto-ID documents, and their availability on the Web was widely publicized.

To explore appropriate responses, the Auto-ID Center established an independent Policy Advisory Council, composed of experts with political, legal, and technology expertise, to help with further research. In addition, the Center formed a technical group, consisting of expert researchers in the fields of privacy, security, and cryptology, as well as a forum for the Chief Privacy Officers from their sponsor companies (Ashton, 2003).

The Auto-ID Center also developed policy guidelines aimed at protecting consumer privacy. These guidelines cover four areas:

1. Consumer notice of the presence of EPC tags on products or in packaging
2. Consumer choice about their options to discard, disable, or remove EPC tags from products they acquire
3. Consumer education about EPC, its applications, and advances in the technology
4. Use, retention, and security of EPC-generated records consistent with all applicable laws

These guidelines were adopted at the final board meeting of the Auto-ID Center in late October 2003, and similar guidelines have since been adopted by its successor organization, EPCglobal.

At the same time, some commercial firms using RFID technology issued their own RFID privacy policies. For example, the Wal-Mart policy, available on its webpage, states that

> Our company is committed to safeguarding the privacy of our customers and members. Wal-Mart and Sam's Club use EPC technology to locate products throughout the supply chain process and inside the stores and clubs. EPC labels do not contain, collect or send any personal information (Wal-Mart, 2009)

Even so, the privacy issue remained a critical flash point. Privacy advocates remain suspicious about the potential of the technologies, while RFID proponents have organized a campaign of education and influence. In April 2004, the California State Senate passed a bill that set limits on the use of RFID technology by libraries, retailers, and other private organizations.[8] Legislators in three other states have announced plans to draft similar bills or to conduct studies.

8.5.5 Release of First EPC Standards

The activities of the Auto-ID Center culminated in the well-orchestrated and widely publicized public release of Version 1.0 of the EPC specifications, also referred to as Generation 1 standards, at the *EPC Executive Symposium*. The symposium, held in Chicago in September 2003, was the first large-scale conference dedicated specifically to EPC technologies. It was intended to help build momentum toward the adoption of RFID across the supply chain. Technical tracks described EPC-compliant technologies and demonstrated reference implementations, while other tracks focused on the case studies and lessons learned from the field trials. At the same time, vendors announced new RFID products and services, including kits targeted to help companies evaluate the technology for their particular applications.

However, Gartner analyst Jeff Woods was skeptical. He believed that the standards announced at the symposium were not mature enough for deployment and that the business case for investment in RFID technologies had not been convincing (Woods and White, 2003). Others questioned whether the network could scale up sufficiently to handle the potentially large volume of simultaneous readings and whether the data collected could be integrated into existing corporate information systems.

8.6 Transition from Research to Commercialization

As early as the fall of 2002, the Board of Overseers reported on initial discussions about the future of the Auto-ID Center. Seeing that the development of the first-generation EPC technology was reaching completion, the Board of Overseers decided that the research and administrative functions should devolve on two different organizations: one focused on continued research and the other on the administration of the standards. MIT would continue to perform research through an organization called Auto-ID Labs. EAN International and the UCC would establish an organization that would be responsible for global, multi-industry adoption of the standards. That transition formally occurred on 26 October 2003, 4 years after the Auto-ID Center was founded. At that time, the administrative functions of the Auto-ID Center at MIT were officially terminated and efforts to develop global RFID standards were transferred to EPCglobal, a not-for-profit joint venture between EAN International and the UCC.

EPCglobal is structured with a Board of Governors, a president, and several action groups (see Figure 8.3). The Board of Governors consists primarily of end users and early adopters of the technology. Business Action Groups comprise representatives from businesses that currently use or plan to use the technology and are

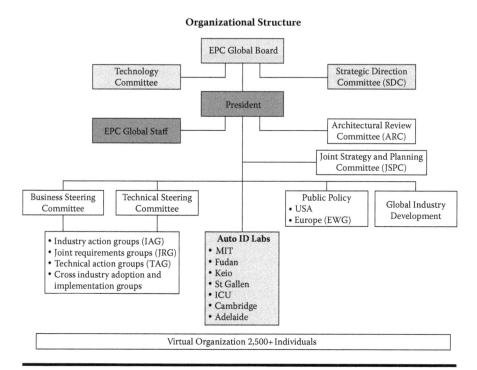

Figure 8.3 The EPCglobal organization. (http://www.epcglobal.org/what/action/group)

chartered to develop business requirements and use cases. At the time of this writing, three such groups are operational (retail supply chain, healthcare and life sciences, and transportation and logistics services) and another four are under consideration (aerospace and defense, consumer electronics, packaging goods, and automotive).

EPCglobal has two technology action groups: one oriented toward hardware and the other toward software. The Hardware Action Group is responsible for defining the standards for hardware components, primarily focusing on the RFID tags and readers, while the Software Action Group is responsible for defining software interfaces and other standards within the EPC network and with other elements of the extended enterprise system. EPCglobal maintains a relationship with Auto-ID Labs, which fosters ongoing research at university laboratories aimed toward proposing new technologies.

In addition, EPCglobal has an architecture review committee, charged with responsibility for creating, documenting, and maintaining the network architecture. The committee is also responsible for identifying components of the architecture that require standards and recommending such standards, as well as for collaborating with other standard-setting bodies.

8.6.1 Early Pilots

Pilots have been the primary means by which early adopters have built a business case, evaluated the technology offerings, and developed their own near-term and longer-term implementation strategies. These pilots have been both internal (focusing on operations within a company) and external (using RFID technologies to link suppliers and their customers).

Companies have used pilots to gain a better understanding of the technology offerings by testing the tags and readers for their particular applications and environments and fine-tuning their implementations. More broadly, these pilots have helped companies develop their longer-term strategies; for example, to identify where process changes may be warranted and where data/system integration may be required, both internally within their corporate information infrastructure and externally with their supplier and retailer partners. While some companies have focused their pilots on complying with their customers' requirements, others have taken more of an enterprisewide perspective, exploring where and how they might implement these technologies across all of their business processes.

Pilots have shown that implementing RFID is not simply a matter of implementing a new technology; it also affects the enterprise's or extended enterprise's applications, infrastructure, business processes, and personnel. One company represented on the Auto-ID Board of Overseers, Kimberly-Clark, has concluded that "successful deployment of RFID will be driven by changes in the business process, NOT technology" (O'Shea and Bigornia, 2004).

8.6.2 Maturation of the EPC Standards

Since the initial release of EPC standards by the Auto-ID Center, EPCglobal has continued to refine the standards.[9] Generation 2 tag protocols for RFID readers in the 860 to 960 MHz ultrahigh frequency (UHF) range have been issued, and a comparable protocol for readers operating in the high frequency (HF) range is under development. In addition, EPCglobal has created specifications for reader protocols and reader management, as well as specifications for *application-level events* (formerly called Savant), EPC information services, and object naming services. Standards for discovery services will follow.

In August 2004, EPCglobal sponsored a series of interoperability tests open to manufacturers of Generation 1 hardware devices. The manufacturers conducted their own tests while representatives of the test facility directed, observed, and recorded the test procedures and results. The tests focused on the ability of different manufacturers to "talk" to each other. Since then, EPCglobal has established a program of conformance tests executed by an independent laboratory. Companies that pass the conformance tests can be certified as compliant with EPCglobal hardware standards.

8.7 EPC Adoption

At an 11 June 2003 joint presentation with the UCC at the *Retail Systems 2003* trade show in Chicago, Wal-Mart's Executive Vice President and Chief Information Officer, Linda Dillman, announced plans to require Wal-Mart's top 100 suppliers to place RFID tags on the pallets and cases they ship to Wal-Mart distribution centers and stores. Wal-Mart planned to "go live" with a limited deployment by January 2005. While Dillman acknowledged that more work was still needed, she commented that Wal-Mart had viewed the Auto-ID Center's field trials as proof-of-concept tests. On the basis of those trials and the results of research that it had performed at its own RFID lab, Wal-Mart expected to start a series of pilot tests that would then be used to design the system implemented in Wal-Mart warehouses worldwide.

Some viewed that announcement as a "line in the sand" (*RFID Journal* staff, 2003). It came 1 month before the smart shelf test was terminated and 3 months before the RFID standards were to be formally introduced. Yet neither event derailed Wal-Mart's commitment to implementing RFID technologies. In fact, in a letter to its suppliers, Wal-Mart announced that it would not only continue with its plan to tag pallets and cases, but also expand the initiative to its next 200 biggest suppliers by January 2006.

Nevertheless, the plan faced substantial concerns and unknowns. Suppliers lacked confidence in the technologies, and questioned whether they were sufficiently mature. They were also concerned about the costs they would incur and whether they would realize any of the benefits internally or whether these would represent merely an added cost of doing business with Wal-Mart. They were also uncertain about what information would be shared between Wal-Mart and its suppliers (Roberti, 2003).

In November 2003, Wal-Mart met with its top suppliers to clarify the RFID requirement and announced a phased rollout starting with three distribution centers and 150 stores. Another 100 centers and 3000 stores would be added by the end of 2005. However, many suppliers remained skeptical. A Forrester Research report published in March 2004 reported that only 25% of the top 100 suppliers expected to meet the 1 January mandate—down significantly from the earlier estimate of 60% (Hines, 2004).

In April 2004, Wal-Mart and eight of its suppliers began a field pilot program in North Texas. Two months later, as reported in testimony before the House Committee on Energy and Commerce (Dillman, 2004), Wal-Mart was tracking cases and pallets of 21 products destined for one distribution center and seven stores.

> At our Sanger, Texas, distribution center, we have placed readers at our receiving doors, above our conveyor belt systems, and at our shipping doors. At the seven Supercenters, we have placed readers at the receiving doors, at strategic points throughout stores' backrooms, at the door to

the sales floor, and at the trash compactor. There are no readers on the sales floor, at the check stands, or at customer entryways or exits. The readers assist Wal-Mart in knowing when a product is received, where it is stored, when it goes out to the sales floor, if it returns for any reason, and when the case is submitted for recycling. This information is shared with our suppliers to assist them with their inventory planning.

Speaking at the annual trade show where a year earlier Wal-Mart's Chief Information Officer had unveiled the chain's RFID plans, Michael Duke, Chief Executive Officer of Wal-Mart's stores division, announced the company's further plans. Wal-Mart intended to expand the RFID effort throughout the United States in 2005 with a goal of involving all suppliers by 2006, and planned on an international rollout in 2005 through 2006.

Additional retailers, including Target and Albertsons in the United States, Tesco in the United Kingdom, and Metro Group in Germany, issued comparable mandates. Target, for example, directed its top suppliers to tag pallets and cases being sent to "select" regional distribution centers beginning in late spring 2005 and expected all suppliers to be tagging pallets and cases by spring 2005 (*RFID Journal* staff, 2004). Metro Group, which had been experimenting with RFID tags and other advanced technologies in its Future Store in Rheinberg, Germany, announced its decision to roll out RFID to 250 additional stores and 10 warehouses (Schwartz, 2004).

Other mandates and policies have contributed to the impetus to implement RFID technology. The DoD issued policy requiring suppliers to use RFID tags on cases and pallets and on individual high-value items for delivery to the military on or after January 2005.[10] In November 2004, the U.S. Food and Drug Administration (FDA) issued special guidance to the pharmaceutical industry, allowing companies to use RFID tags without violating regulations that govern product labeling.[11] The FDA also published guidance for using RFID technology in feasibility studies and pilot programs designed to increase the safety and security of the nation's drug supply. The initiative was part of an ongoing FDA effort to combat the rise in counterfeit drugs.[12] Immediately thereafter, three pharmaceutical companies announced that they would start tagging shipments of drugs that are most subject to counterfeiting.

In January 2005, Wal-Mart reported that RFID was operational in about 140 stores and 3 distribution centers, and that the technology was yielding benefits not only in inventory, but also in visibility into the efficiencies of Wal-Mart's supply chain. By early 2007, however, Wal-Mart acknowledged that it had missed its goal of installing RFID technology in 12 of its 137 distribution centers and reported that this reflected a change in its strategy from one focused on its distribution centers to one focused on implementing RFID in selected retail stores (Songini, 2007). This shift in strategy suggests that, notwithstanding Wal-Mart's mandates, widespread adoption has been slower than anticipated. And because the market did not grow as expected, the costs of tags and readers, while lower than at the outset, had not reached the 5-cent-per-tag level.

8.8 Summary

RFID has the potential to change the global supply chain significantly. Since its origins in 1999, it has moved from an academic research initiative to initial rollout by a small, but highly influential, group of early adopters. It has sparked controversy but that controversy has neither deterred nor derailed implementation. It has, however, caused proponents to address issues of consumer privacy directly and to focus their efforts on back-of-the-store applications that do not affect consumers directly and thus may have delayed widespread deployment of RFID technology.

Development and well-orchestrated marketing of the technologies have occurred collaboratively and in parallel. Adoption of the RFID technologies has been facilitated by the development of open, global standards and was further advanced by the advocacy of large retailers, such as Wal-Mart, that have the clout to influence the actions of their suppliers. More companies are conducting or planning pilots not only to comply with mandates from their customers, but also to help define the business case for their own applications. These pilots have proved critical in helping companies and their partners determine how to implement the technologies and how these technologies can help their bottom line, and to identify what processes must change in order to leverage the full potential of RFID technologies. These companies are, in essence, learning by doing.

However, some analysts and practitioners acknowledge that significant hurdles must be overcome before RFID can realize the lofty goals of its advocates. To fulfill its promise, RFID depends on continuing improvements to the technologies, reduced costs, and the ability to work across different generations of the technology. But perhaps the most significant hurdles come not from the technology, but from the need to change underlying business processes in order to leverage the data generated by this technology. Kimberly-Clark saw a continuum of increasing functionality and increasing value:

■ *Phase I: Compliance.* Companies are only concerned with complying with mandates. They invest in encoding and tagging their products and record the data, but do not use the data for any internal business processes. This phase has also been called "slap and ship."
■ *Phase II: Collaboration and Visibility.* Companies are sharing data with their strategic partners and are able to track and trace goods across the supply chain. However, other business processes are not affected.

- *Phase III: Zero Latency Decision Support.* Data is collected and analyzed in real-time. For example, RFID could be used to provide electronic proof of delivery, thereby reducing invoice discrepancies and improving the visibility of transactions.
- *Phase IV: RFID Enabled Adaptive Enterprise.* RFID data is fully integrated with existing business processes and IT systems. Real-time data from RFID technologies can be used to trigger business events and drive faster business decisions, in effect providing a "sense-and-respond" capability (O'Shea and Bigornia, 2004).

When this presentation was given in 2004, Kimberly-Clark viewed industry as being somewhere between Phases I and II. To advance to the higher phases, the company anticipated that efforts must go beyond merely implementing the technology or even continuing to improve it, to integrating it in the business processes and using the data to drive better business decisions. To make that happen, they saw the need to integrate internal data and systems, to synchronize data, and to improve customer and supplier collaboration and communications.

By 2007 it still appeared that while the technology had stabilized and some companies were reaping business value by reducing out-of-stock levels, the EPC market had not grown at the rates predicted earlier. Consequently, the prices of tags and readers have declined, but not to the levels originally anticipated. Widespread adoption is still some years away, and it is likely that the industry overall remains somewhere between Phases I and II.

8.9 EPCglobal Network Mapping to the Extended Framework

The vision that spawned the EPCglobal network was based on leveraging Internet technologies to track items across the open, global supply chain. Therefore, this case study focuses on engineering in the context of the extended enterprise. While the technology was not new, its application in this business sector was. The scope of the effort all items from manufacture to sale to disposal—was audacious. The approach to executing it, using a consortium of universities to conduct the research and a consortium of natural competitors to fund it, was also remarkably innovative. It is not surprising, then, that its profile shows performance in all four quadrants in or near the outer ring (Figure 8.4).

From the *mission environment* perspective—in this case, the global supply chain—the pace of change has increased with the expansion of the Internet and information technologies as well as with the globalization of markets. It is expected—or hoped—that the introduction of RFID technologies will not only respond to the changing needs of the supply chain, but also further change it,

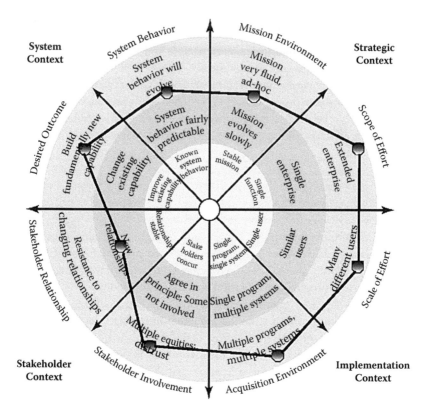

Figure 8.4 The EPC profile. (From Stevens, R., 2008. *Profiling Complex Systems*, Proceedings of the IEEE International Systems Conference, Montreal, Canada, April 2008. With permission.)

possibly quite dramatically. In this case, the changing needs of the environment drive the vision for the technology; in turn, implementing the technology promises to change the environment. The two co-evolve.

This case represents a classic example of an extended enterprise and was selected for that reason. In terms of the *scale of effort*, it crosses multiple corporations, including suppliers of consumer products as well as retailers, and allows for data sharing across corporate boundaries. However, it is worth noting that the interests of the different members of the supply chain can and do differ. Retailers are primarily interested in reducing out-of-stock situations while suppliers are primarily interested in improving inventory management and supply chain visibility. While these interests are not mutually exclusive, some suppliers have voiced the concern that implementing RFID benefits the retailer, but is only an additional cost of doing business for the supplier.

In terms of the *scale of the effort*, RFID proponents expect the technologies to be used by a wide range of industries and business sectors and at different stages in the supply chain, from initial manufacturing through distribution to retail operations. RFID can allow companies to track high-value assets and to distinguish between authentic and counterfeit products, among others. Different implementations will be needed for these different market segments and uses.

From an *acquisition* perspective, there are many vendors of EPC-compliant products and services. The commitment to a set of open standards allows these different implementations to standardize on common data structures and means of capturing and exchanging EPC data.

From the *stakeholder* context, the broad support for this development effort is noteworthy. That natural competitors were willing to cooperate in funding and supporting EPC developments reflects their recognition that a common solution was in everyone's best interest. However, at the same time, it is worth noting that these were not the only groups that had a stake in the matter. The EPC consortia did not include consumers and privacy interest groups. Instead, in some cases, EPC advocates viewed them primarily as opponents to manage or neutralize, but in fact their protests led to changed plans and strategies. Moreover, not all suppliers support the introduction of RFID, and many that have been directed to implement the technologies have done so only reluctantly.

From the perspective of the *desired outcome*, this effort evidently set out to address some well-recognized problems in the supply chain by building a fundamentally new capability. The "new" elements are not so much the technology that replaces (or augments) the bar code, although its application was certainly novel, but rather the implications and opportunities that the technology offers for changing some of the underlying business processes and organizational relationships.

Finally, from the perspective of *system behavior*, it is difficult to separate the behavior of the technologies themselves from the way they are used in business processes. Early field trials proved particularly useful in helping the technology developers understand both the opportunities and constraints of the particular business environments. Similarly, while proponents expected that early pilots would be used primarily to fine-tune the implementation of technologies, such pilots continue to be used to help define the business process and to further refine the value proposition. In effect, these pilots allow users to experiment with the best approaches for implementing RFID technologies.

8.10 Insights for Engineering Mega-Systems

■ *Stakeholders.* The greater the number of potential stakeholders, the greater the value of establishing and nurturing consortia that represent their interests.

It is also vital to include all stakeholders: not only advocates, but also, more particularly, skeptics and potential opponents.

■ *Novel projects.* Novel projects often demand novel organizational arrangements and processes.

■ *Early field trials.* Early and incremental field trials provide a valuable venue for gaining insight into not only the actual performance of the technologies in situ, but also into real-world implementation. The feedback gained from such events can be used to refine requirements and concepts of operations.

■ *Requirements assessment.* Particularly in situations where the underlying business case and concept of operations are still evolving, it is possible—and probably highly desirable—to evolve the requirements concurrently with the technologies themselves.

■ *Vision.* A well-articulated, agreed-to vision, coupled with a few key driving requirements, can serve as a sufficient basis upon which to build an enterprisewide capability.

■ *Decisions in a broader context.* It is important to understand not only the technical issues being addressed, but also their political, operational, and economic implications. Failure to do so risks the possibility that the project, while technically feasible, loses support and hence viability. Attempting to resolve political issues by purely technical means generally proves inadequate.

■ *Advocacy.* Advocacy of market leaders can jump-start the adoption of common technologies. However, widespread adoption depends on many individual decisions, each focused on considerations of local costs and benefits.

■ *Pilots.* Pilots, and other forms of operational experiments, are valuable in helping users understand the potential of the technologies being implemented and their potential impact on local business processes.

■ *Value.* The value of the system being implemented is often found in the changes that it sparks in the underlying business processes. Many of these changes cannot be anticipated, but will evolve over time.

Endnotes

1. These are the standards development organizations in the United States and Europe that are responsible for setting and administering bar code standards.
2. While RFID technologies had been developed over a number of decades, it was not until the 1980s that there were widespread commercial applications as diverse as animal tracking, vehicle tracking, factory tracking, and personnel access. RFID for collecting tolls was introduced in 1987 in Europe and 2 years later in the United States. By the late 1990s, electronic toll collection was widespread and RFID was being used for point-of-sale (POS) applications for fuels and fast foods (Landt, 2001).
3. "Shrinkage" is the supply chain term of art that covers inventory loss, misplacement, and theft.

4. Auto-ID Labs. Undated. *History of the Auto-ID Center.* Cambridge, MA: Auto-ID Labs.
5. The DoD Architecture Framework (DoDAF) defines three architecture views: operational, systems, and technical. The operational view describes the tasks and activities, operational elements, and associated information flows. The systems view provides a description of the systems and interconnections that support the operational functions. The technical view defines the criteria that govern each required system capability.
6. For example, they learned how high and how wide to stack the goods for optimum reading results.
7. As an anti-theft measure, the smart shelf tracks the number of packages of razor blades being removed. If a consumer picks up three or more packs of razor blades, the system audibly thanks the consumer and automatically alerts security to a potential theft. The expectation is that consumers will like being thanked and potential thieves will realize that they have been detected and so will not commit the theft. The system is also capable of taking a photograph of the suspected thief for comparison with photographs taken at checkout (Texas Instruments, 2003).
8. The bill would prohibit businesses and libraries in California from using tags attached to consumer products or using an RFID reader that could be used to identify an individual unless the technologies comply with certain conditions. Specifically, information can only be collected for items that customers are actually buying, renting, or borrowing. Information cannot be collected on what customers may have picked up and put back, what they are wearing, or what they may be carrying in a wallet or purse (Swedberg, 2004).
9. These standards cover tag data specification, hardware, software, and information services. Tag standards have been expanded to support several different coding schemes, in addition to the original code developed by the Auto-ID Center. RFID protocols address the "over-the-air" interface between the tag and the reader.
10. Information about the DoD RFID policy is available at: http://www.acq.osd.mil/log/ rfid/rfid_policy.htm (accessed 24 February 2008).
11. U.S. Department of Health and Human Services, Food and Drug Administration, *Radiofrequency Identification Feasibility Studies and Pilot Programs for Drugs,* Guidance for FDA Staff and Industry, Compliance Policy Guides, Sec. 400.210, November 2004. http://www.fda.gov/oc/initiatives/counterfeit/rfid_cpg.html.
12. The U.S. Food and Drug Administration Compliance Policy Guide is available at www.fda.gov/oc/initiatives/counterfeit/rfid_cpg.html.

Chapter 9

Observations from the Case Studies

While the case studies of SIAP and RFID have limitations for all the reasons mentioned earlier, they do provide us with insights from experiences in the real world. On the basis of these experiences, we can begin to formulate some ideas about appropriate principles, practices, and tools for engineering mega-systems.

9.1 Case Study Recap

At the outset, it is important to note some key differences between the two examples of mega-systems and traditional systems engineering projects. In both cases presented, the systems engineers are developing what is, in essence, a horizontal, cross-cutting capability that must then be integrated into individual warfighting systems (in the case of

> Engineering mega-systems require the system engineer to work across multiple, independent systems to define the elements that ought to be in common. And he or she does this often without the authority to impose compliance on the "receiving" systems.

SIAP) or into the business processes of individual companies and market sectors (in the case of RFID). This division between specification and subsequent integration is not typical of most systems engineering projects, which retain responsibility over the system's entire life cycle, but it appears more typical of the kinds of mega-systems projects we have described. In these cases, the systems engineer works across multiple, independent systems and does so by defining the elements that these systems should have in common, yet often lacks the authority to impose compliance on the receiving systems. This is a critical difference between systems engineering of traditional well-bounded projects and systems engineering of mega-systems.

In both the cases studied, the effort was directed at developing and evolving specifications for these common elements. In the SIAP case, these specifications were developed as a behavioral model instantiated in code. That model is then delivered to a number of separate programs for each to integrate into its own warfighting systems. In the RFID case, the effort was initially directed at developing the first generation of standards for hardware and software. Developers continue to mature these standards with the expectation that they will be commercialized by a number of separate, competing vendors. Thus, unlike more traditional efforts where the systems engineer is responsible for specification of the product and its components and then assumes responsibility for the integration of these components into the overall system, in these cases, the systems engineering effort is only responsible for the left leg of the "V" Model (see Figure 5.1). Other actors assume responsibility for not only developing the products, but also integrating them into the larger systems in which they will operate.

Tables 9.1 and 9.2 provide an overview of the case studies. Table 9.1 describes the context in terms of the programs' objectives, organization, and external influences. Table 9.2 then synthesizes the approaches and activities used. Its organization is consistent with the Friedman–Sage structure (Friedman and Sage, 2004) and focuses on the six concept areas that represent phases in the systems engineering life cycle.

9.2 Observations

The mega-systems discussed in the case studies fall into the category of "enterprise networks" that enable enterprise-wide and even extended enterprise-wide operations. The term "enable" is important because these networks themselves do not constitute the end-state, but rather the means to achieve the desired organizational or trans-organizational objectives. In the case of SIAP, the desired enabler was a standardized air picture that could be shared by warfighting systems built by different Services. Such a capability would make a key contribution to an integrated theater-wide air and missile defense capability. In the case of RFID, the desired enabler is visibility of products throughout their entire life cycle—from initial manufacture, through distribution, sale, and ultimately disposal. The ultimate objective is not just to gain a positive return on investment through reduced costs and increased returns, but also to change the business process fundamentally by moving toward what has been described as an adaptive enterprise—an enterprise in which supply and demand are synchronized in real-time.

Table 9.1 Overview and Comparison of Case Studies: Context

	SIAP	*RFID*
Objective	Develop technical solutions that could be applied to multiple systems to yield consistent and accurate representations of air tracks. Eliminating unwanted behavior is key.	Develop and implement technologies to improve (and in some cases, fundamentally change) how items are tracked throughout the global supply chain.
Organization	Initially organized as a Systems Engineering Task Force within the U.S. DoD. Subsequently reorganized as a more formal Joint SIAP Systems Engineering Organization and, more recently, as a Joint Program Office. Initially structured as a "virtual team" with service stakeholders. Since then, has evolved into a consortium that includes representatives from the developers of the targeted systems (SIAP's customers).	Initially established as a university-based research center with a mission to develop and promote open Auto-ID standards; led by a former industry product manager. Research funded by a broad-based consortium of industry and standards organizations. Later transitioned to a not-for-profit joint venture responsible for global, multi-industry adoption of the standards.
External Influences	Despite general agreement about the objectives of the effort, there have been recurring stakeholder concerns about process, products, and impact on performance of the receiving weapon systems. The program was formally terminated as of September 30, 2009.	Privacy concerns impeded original vision to tag individual items and forced early adopters to redirect initial efforts to back-of-the-store applications.

Because these mega-systems are often intended to address strategic objectives that extend beyond the specifics of their design or technology, it becomes necessary to understand the broader context in which they are developed and evolve. For this reason, these case studies have explored not only traditional systems engineering processes, but also the political, organizational, operational, and economic factors that shaped these efforts, either by driving or constraining them.

Table 9.2   Overview and Comparison of Case Studies: Approaches and Activities

	SIAP	RFID
Requirements Definition and Management	SIAP-related requirements extracted from multiple sources and used to derive system requirements for initial implementation. Capabilities Development Document formally approved for a CD-1 in late 2007.	No formal requirements, only a vision and a concept. Developed the justification (business case) in parallel with the technologies. Focused efforts on consumer products domain because that was where initial interest lay and where industry recognized a need.
Systems Architecture Development	Developed static architecture products (views) for IABM consistent with DoD framework. Operational threads used to describe specific sequence of activities and provide operational context. Extended architecture effort developing dynamic behavioral models.	No formal architecture products per se. Developed professional graphic products showing envisioned future, intended for marketing and education purposes.
System, Subsystem Design	The initial engineering process was subsequently extended to consolidate design guidance for IABM development and to maintain consistency among requirements, architecture, and design.	Need for low-cost tag and ubiquitous deployment drove the initial design, which mimicked the Internet model. Emerging business needs drove the design of next-generation standards.
Validation and Verification	Using a federated hardware-in-the-loop and digital simulation environment to assist in development and integration testing of incremental builds. Plan is to use similar environment to test each of the Platform Specific Models, first individually and then in combination. The Air Force-established Integration Resource Center is used to conduct independent verification and validation.	Early field trials yielded many lessons from the real world. Companies initiating pilots to tailor applications to local needs and simultaneously build the business case. Test sites set up to demonstrate interoperability across vendor products. Plans in place to conduct compliance testing.

Table 9.2 (continued)   Overview and Comparison of Case Studies: Approaches and Activities

	SIAP	RFID
Systems Integration and Interfaces	Each of the targeted programs was to be responsible for integrating the executable code into its systems. This was a recognized risk and a potential area for unexpected system behavior. Recognizing this, there was interest in identifying ways to reduce the level of complexity.	Original activities did not consider integration into existing corporate enterprise systems. Since then, vendors are offering applications that integrate RFID data with existing enterprise applications data.
Deployment, Roll-out	Capability Drop 1 was released in late 2008. Funding shortfalls hampered efforts to integrate the release into individual weapon systems. The program was formally terminated at the end of FY2009.	Mandates (Wal-Mart and others) have spurred implementation. Maturity and cost of the technologies still raise questions and widespread adoption is still some years away.

Mega-systems are, in fact, being engineered and developed through the cooperative and collaborative behavior of large enterprises, including government agencies, businesses, and global joint ventures. We expect to see more of these types of systems and more examples of such cooperation and collaborative developments.

Our examinations of an admittedly limited set of mega-systems still allow us to synthesize observations not only about the differences between traditional systems engineering and the engineering of these large-scale, cross-boundary systems, but also about the principles and practices that seem most effective. We have derived the following observations from these specific case studies. It would be valuable to explore the extent to which they continue to apply to other examples of mega-systems.

9.2.1 Greater Diversity of Stakeholders and Interests

The two case studies demonstrate the critical role that stakeholders play in both advancing and, in some cases, constraining the engineering and deployment of these mega-systems. The broader the scope of the effort, the greater the likely diversity of the individuals, organizations, and interests that are involved or have an interest in the project. Consequently, the likelihood increases that these interests may conflict. Ignoring some key stakeholders or underestimating their ability to influence the outcome poses great risks to the project.

Both SIAP and RFID have organizational models that were deliberately designed to bring together key stakeholders. SIAP formed a consortium of developers and funded the contractors who are building their individual targeted systems (in effect, their customers' contractors) to participate in the development process. Similarly, RFID standards were developed by a global consortium of universities and were funded by a consortium of businesses, many of which continue to be direct competitors.

Such consortia have a critical role in forging a common perspective and building trust among the various stakeholders. Consequently, it is important to involve all stakeholders, particularly those with an operational (rather than a purely technical) stake in the outcome. It is worth noting that the RFID consortium did not include consumers as stakeholders, perhaps because they were not viewed as having an economic interest in the technology's development or application. One can speculate that if RFID proponents had addressed consumer interests earlier, the interactions around privacy concerns might have been less adversarial and contentious. On the other hand, the SIAP consortium brought together the right stakeholders, but involvement in the development process did not necessarily translate into a commitment to integrate SIAP into the separate weapon systems. In other words, involvement, by itself, does not necessarily lead to commitment, and verbal commitment does not necessarily translate into necessary actions.

9.2.2 Broader Set of Considerations

Ever since its emergence as a separate discipline, traditional systems engineering has balanced technical and cost considerations in conducting trades and selecting feasible design options. Mega-systems engineering must obviously continue to address these dimensions, but must also encompass a richer set of considerations, including political, operational, organizational, and economic factors. In some cases, mega-systems engineering must also take cultural factors into account, not only because of the realities of global partnerships, but also because different organizations have their own unique cultures, with different assumptions, values, modes of operation, and even vocabularies.

Thus, for mega-systems engineering, focusing primarily on the technical dimensions is not enough: The best technical decisions may turn out not to be the most viable decisions. The technical approach may prove too costly or may lack the support of key stakeholders. The sheer scope of mega-systems increases the probability that the external environment may change over the course of the project, and what appeared to be a reasonable solution at one time may no longer be as appropriate. This, of course, can also occur in the course of engineering more bounded but primarily deterministic systems, but the challenge becomes significantly more complex when engineering large-scale mega-systems. Not only do such systems involve more diverse organizations, each with its own interests, but the transformational nature of many of these mega-systems also introduces public policy issues that systems engineers cannot ignore and must address.

In the SIAP case, political, organizational, and economic aspects had a significant impact on the organization's ability to proceed. These forces drove changes in the SIAP program's structure, relationships with affected programs, and the way the SIAP capability was developed and implemented.

In the RFID case, the unique organizational structure proved effective in bridging three very different cultures: (1) that of a research organization housed in a university, (2) a business organization charged with making a profit, and (3) a standards development organization. However, although the consortium acknowledged the issue of consumer privacy early on, it failed to deal adequately with the associated concerns. These concerns were clearly emotional and often focused on worst-case possibilities, but the consortium's response had initially emphasized technical solutions (e.g., adding a "kill" command to the specification). Later on, the consortium attempted to counter the concerns of privacy advocates through a media-like campaign but that strategy, by all accounts, was turned against the RFID project.

9.2.3 Convergence on Critical Infrastructure Standards and Design Tenets

The larger the number of different systems and organizations involved, the greater the importance of converging on infrastructure standards and design tenets. The field of systems engineering certainly does not lack standards. On the contrary, systems engineers have many standards from which they can choose; and even when a particular standard has been mandated, it can still allow options or be open to interpretation. In such instances, the value that would accrue from complying with a standard disappears, as demonstrated by the incompatibilities that precipitated the establishment of the SIAP effort. The challenge to systems engineers is less about selecting all the standards to govern their systems and more about determining the minimum set of standards[1] that should be followed and that require unwavering compliance. In fact, the smaller the set of standards, the greater the likelihood that the standards will be followed.

The RFID case provides an example of the willingness of businesses, including direct competitors, to converge on a common set of open standards and then remain prepared, collectively, to modify those standards as business needs warranted. Admittedly, because the EPC standards leveraged widely used and well-recognized Internet and World Wide Web standards, they had a ready foundation on which to proceed.

Selecting the appropriate design tenets is a related but somewhat different issue. Consider a case where the participating elements—be they organizations, nodes on a network, or individual systems—are known a priori and their behaviors are prescribed. In such cases it would make sense to have a tightly coupled design that seeks to gain maximum efficiencies. Now consider the opposite case, involving a mix of anticipated and, more importantly, unanticipated participants, and the nature of their interactions is not fully predictable. In these cases, a more loosely coupled design approach relying on a minimum set of agreed-to standards allows for the necessary flexibility and adaptability. By doing so, however, it may sacrifice some measure of efficiency. A mixed or hybrid approach is also feasible, where individual systems are tightly coupled internally but loosely coupled in their external interactions. Selecting the design pattern that best suits the degree of uncertainty in the underlying concept of operations is a key consideration in mega-systems engineering.

9.2.4 Requirements

Traditional systems engineering, like software engineering using the Waterfall Model, is predicated on having a well-defined and precise set of requirements that remain more or less stable over time. The types of mega-systems that we have explored are both more complex and interdependent than traditional systems, and are expected to be more adaptable to meeting changing needs and expectations. At the same time, much of the underlying technology is also changing rapidly, offering opportunities for new functionality that was likely not envisioned when the effort was initiated. Thus, for many mega-systems, it will be unrealistic to expect to bound and control requirements in the same manner that applies to more deterministic systems. Instead, it would be best to articulate requirements initially as broad vision statements, concepts of operations, or architectures, with the expectation that they will necessarily evolve over time in response to changing needs, opportunities, and constraints.

The SIAP program had no formal requirements available at the onset. Instead, de-facto requirements were derived from various official documents to provide a measure of the legitimacy demanded in the DoD environment. This did not prove adequate for such a complex undertaking, and the community tried to converge on more definitive requirements, with an initial focus on the first increment of functionality. As the SIAP organization transitioned to a more formal acquisition organization, the requirements documentation and approval process also became more formal.

RFID, because it operates primarily in a commercial business framework, did not need a comparable set of formal requirements. A top-level, simply stated vision apparently sufficed to drive the design. Participation of a knowledgeable end user, in this case Kevin Ashton, Director of the Auto-ID Center, meant that the developers had direct access to a source of expertise that enabled them to understand the needs of the customer, and that these customer requirements did not have to be filtered through a formal process or intermediary organizations. Interestingly enough, RFID did actually develop requirements, in the form of business cases, but did so in parallel with the development of the technologies and more for marketing purposes than to form the basis for developing a design or specification.

RFID requirements have continued to evolve. As participating companies have begun to gain experience with the technologies, they have recognized additional needs that had not been fully anticipated at the outset.

9.2.5 Discovery Engineering

The greater the uncertainty in the initial requirements, the greater the importance of all types of methods and tools that allow for the exploration and understanding of required system behavior and the evolution of system features. These methods and tools encompass a wide range of activities, including early prototyping, exploratory integration, modeling and simulation, field trials, pilots, and experiments. Collectively, these activities can be considered a process of discovery engineering.

Both SIAP and RFID were willing to expose not fully mature or complete products to their customer base to gain early feedback. In addition, they were willing to use that feedback both to refine requirements and to gain an understanding of how the systems they were specifying would operate in the larger environment.

In the case of SIAP, integration of its software into target warfighting systems must wait until the SIAP computerized specification is relatively complete and stable. This occurs only after a period of integration and validation in a laboratory environment. That process is expensive and requires careful consideration of the ways in which the SIAP computerized specification will affect the behavior and performance of the warfighting system itself. Hence, a simulated environment offers the best setting for systems engineers to gain early insight into the behavior of the SIAP-generated code both within a single platform and then between and among platforms. SIAP had, from the onset, emphasized exploration, integration, and test using such simulated environments, both in single laboratories and then distributed among multiple sites.

RFID has relied on early prototypes, early field trials, and pilots, not only to evaluate and refine the technologies, but also to gain insights into real-world implementations. Even after the underlying specifications have been agreed to and the technology has become available commercially, a pattern of extended pilot activities leading to staged roll-out continues. Many of the more recent piloting efforts—in effect, real-world experiments—are being conducted by companies that are exploring RFID technologies for their business practices and are determining specific uses that can reduce costs, increase profits, or guarantee authenticity (as in the case of the pharmaceutical industry).

9.2.6 Role of Grand Design

Grand vision need not equate to a grand design. Changing circumstances, including changes in user expectations, will necessitate changes to the initial design. The greater the novelty of the effort, the degree of uncertainty in the broader receiving environment, or the diversity among the stakeholders, the more important it is to conceive a design that can accommodate such changes.

SIAP used a model-driven architecture approach that links analysis, design, test, and the software code itself. The behavioral model is implemented in a series of increments, known as time boxes, each of which allows for feedback to the evolving design. The SIAP program expected to periodically release capability drops that provide a sufficiently mature level of functionality so that it can then be integrated into warfighting system platforms.

RFID developed its initial design on the basis of a relatively unconstrained view of the environments in which it was to be implemented. At a top level, the basic design originally conceived by MIT researchers remains valid. However, the specifics have continued to evolve, reflecting feedback not only from early trials and ongoing pilots, but also from strong proponents and interested skeptics.

9.2.7 Managing Risk and Uncertainty

Risk management is a well-recognized component of project management and systems engineering. It is an established discipline with formalized, well-documented processes and acknowledged best practices. While most risk management experts agree that risks can be either positive or negative, in practice most practitioners view risks as having primarily negative impacts on project outcomes. Because of the complexity and uncertainty inherent in many mega-system projects, as well as their diverse constituents, mega-system engineering must be prepared not only to consider downside risks, but also to leverage unanticipated opportunities (Hillson, 2004) (see discussion in Chapter 11).

In the SIAP case, the technical and project leadership recognized that the initial strategy of incremental block releases was not working as expected and changed the fundamental approach of the effort. Rather than focus on fixing existing specifications, the new approach centered on developing what is, in effect, common software. In doing so, SIAP leveraged an emerging design and development process and toolset.

In the RFID case, the original design was driven by the need to track individual items and the associated scaling requirements. When privacy advocates raised concerns, the emphasis shifted from tracking individual items to less controversial back-of-the-store applications. At the same time, RFID introduced new item-level applications that yield more visible consumer benefits, such as tracking pharmaceuticals to combat drug counterfeiting.

9.2.8 Leadership

The nature and quality of the leadership in a mega-system project are particularly important in building and sustaining external support, recognizing changes in the external environment, and adapting to these changing circumstances. In some instances, leaders must have the ability to recognize opportunities and develop appropriate strategies to leverage them better. In other cases, they must be able to devise solutions to previously unanticipated problems. In the traditional systems engineering process, these qualities are brought to bear primarily, although clearly not exclusively, during the initial stages of the effort. In the engineering and evolution of mega-systems, these skills and talents can be called for throughout the effort.

The SIAP leadership had, for much of the history of the program, been remarkably effective in gaining senior leadership support and in leveraging that support to bring initially reluctant stakeholders into the consortium. Their flexible outlook also allowed them to recognize that the initial block approach was unlikely to achieve expected SIAP outcomes and to propose a drastic course change.

In the RFID case, the Auto-ID Center drew its executive director from the industry that proposed to act as the first implementer of the technologies. This gave him, and the research team he led, a clear advantage in terms of his ability to understand and interact with the companies funding the effort.

Endnotes

1. Consider the example of the Universal Core (UCORE), an effort intended to make it easier to share information across and within communities of interest. Rather than trying to standardize formatting of all the possible data, this effort focuses on standardizing a small set of universally understood concepts (where, when, who). Phase I addressed information sharing between the DoD and the Intelligence Community. Phase II expanded the effort to include the Department of Justice and the Department of Homeland Security.

THE WAY AHEAD

Chapter 10

The Way Ahead

As noted in the introduction, this book is not intended as a definitive work, but rather as an early step toward the creation of a body of knowledge about the engineering, development, and evolution of the large-scale systems that we term "mega-systems." We hope it serves to foster and contribute to a nascent but expanding dialogue among practitioners, academicians, researchers, and customers. With this goal in mind, we have sought to define and frame the concepts and practicalities of engineering mega-systems, examine some of the dimensions, and develop a framework to help us explore them further. Just as traditional systems engineering evolved through practice in the post-World War II era, we fully expect that the engineering of this class of systems will also evolve through practice. In this way the community will collaboratively generate a body of precepts, practices, and even tools that will prove useful in the engineering, development, and evolution of this class of systems. This chapter synthesizes the emerging systems engineering tenets and suggests a way ahead.

10.1 Emerging Tenets

While the case studies presented have acknowledged limitations, we nevertheless believe that we can begin to generalize the observations into a set of emerging tenets, or principles, related to the engineering of these mega-systems. Further, we believe that in mapping these emerging tenets to the Profiler, we can create the beginnings of a situational model. By that, we mean that the practices and processes most suitable for a given system or mega-system depend on the particular situation at hand. What works for a well-bounded system will not necessarily work for a system that is intended to bridge multiple organizations. Similarly, what works for a system with a well-understood mission and well-defined and stable business processes will not necessarily work for a system that is expected to operate in a rapidly changing environment. The key, we believe, is first to understand the circumstances that apply to the particular effort at hand and then, on the basis of that assessment, to pick the most suitable set of tools and techniques. One size definitely does not fit all!

Figure 7.5 (Chapter 7) and Figure 8.4 (Chapter 8) illustrated how the Profiler first introduced in Chapter 5 could be used to build a polar diagram specific to each of the case studies. These polar diagrams depict the situation particular to that mega-system, recognizing that any such characterization would necessarily change over time as the circumstances that it captures themselves change. Figures 10.1 through 10.4 step through the Profiler, highlighting each of the four quadrants of the Profiler in turn. For each quadrant we examine where differences in emphasis from traditional systems engineering may be required. Following that, we discuss the upper and lower hemispheres of the Profiler, with an emphasis on where the particular characteristics of a mega-system may warrant extensions to traditional systems engineering practices.

10.1.1 Strategic Context: Iterative and Incremental Strategy

The features of the strategic context (Figure 10.1), characterized by the degree of flux expected in both the mission environment and the scope of the effort, will influence the overall development and deployment strategy. If the mission environment in which the system of interest is expected to operate is changing rapidly—either in terms of the participants and their interactions or in terms of the underlying business practices—then it is likely that user needs and expectations will also evolve and so, in turn, will the requirements for the system(s). Under such circumstances, it would be counterproductive to lock in requirements at the beginning and expect them to remain valid and unchanged throughout the duration of the effort. Rather, the systems engineer should expect and be prepared to accommodate evolving requirements rather than resisting them. It would be appropriate to adopt an iterative strategy that allows for smaller increments of delivered capability and the opportunity to obtain critical user feedback and adapt the next iteration accordingly. In fact, the more volatile the environment, the more frequent the iterations should be. Such a strategy would allow the program to address unanticipated user and not-yet-defined features and to accommodate emerging technologies.

This strategy should emphasize *spiral fieldings*. Note that these are not just spiral developments, but in fact spiral drops that provide some incremental set of actual capability to the targeted users. In other words, they have a market value. Experience with successive increments allows users to refine their evolving needs and provide *feedback* to the developers, while at the same time accommodating the inevitable changes in operations, technologies, and user expectations. This is quite similar to the commercial model used for new product development. In that model, while the outcome space is known, the exact form of the final product may not be.

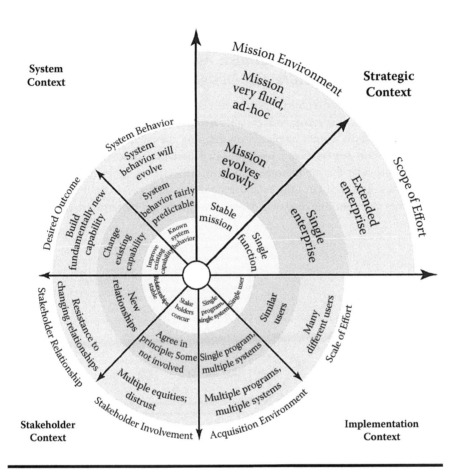

Figure 10.1   The strategic context.

The new product developer progresses via a series of versions, each building on the previous one by improving on existing functionality or adding new features, and at each step reflecting the developer's insights into the emerging market. In contrast, the more stable the mission environment, the more likely it is that the requirements will remain valid over time. Under those circumstances, a more traditional top-down systems engineering strategy is appropriate.

In situations where the system is intended to cross multiple boundaries—for instance, a system intended to work across an enterprise or to link strategic partners in an extended enterprise (particularly when these partners lack a history of working effectively with one another)—there is merit in structuring the acquisition strategy to focus initially on pilot activities. Such pilot activities would address a selected slice of the overall effort and would be directed as much toward building trust as toward addressing substantive issues of terminology, operational patterns, and desired features.

10.1.2 Implementation Context: Agreement on Infrastructure and Design Tenets

In the implementation context (Figure 10.2), the larger the number of separately managed systems that must work collaboratively to provide the needed capability, the more emphasis should be placed on defining the enabling infrastructure, the common design patterns, and recommended best practices. The enabling infrastructure provides for the interoperability of information, products, and technology services, and facilitates the establishment of a common basis from which the capabilities can continue to evolve over time. The design patterns are the essential

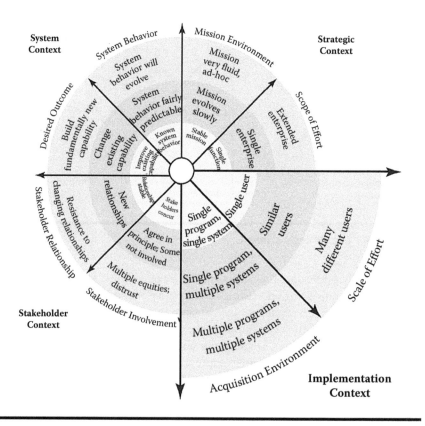

Figure 10.2   The implementation context.

principles that guide how the different systems are architected and built. For example, situations in which there are multiple legacy systems *and* where the nature of the interactions between them is difficult to anticipate completely would be best served by design patterns that emphasize flexibility and adaptability. Loose coupling is such an approach to designing resilient interfaces. It limits interdependencies among components and is intended to reduce the risk that changes in one component will yield unanticipated changes in others. In contrast, situations that depend on high levels of synchronization would be best suited to design patterns that emphasize tight coupling.

The simpler and leaner the set of infrastructure standards and design tenets, the more likely it is that the separate programs will be able to reach consensus around them. This is, in effect, the structured part of what may be a very unstructured problem.

In contrast, a single program that is not expected to involve interconnection or interdependence with other systems or programs can define its own infrastructure. It is, in effect, a closed system and can operate independently.

10.1.3 Stakeholder Context: Working to Identify Intersecting Interests

In the stakeholder context (Figure 10.3), the greater the degree of diversity among the key stakeholders, the more important it is that engineers and program managers understand the positions of each of the key stakeholders and actively work to identify areas of potential intersection. Techniques such as stakeholder analysis become more critical. Bringing stakeholders into the process—for instance by engaging them in trade-off analyses—offers opportunities to develop mutually acceptable strategies. In cases where there are divergent stakeholder positions, it will be impossible to meet all of their separate requirements, but it would be critical to identify the subset where their needs intersect and establish that as the priority effort.

The greater the number of stakeholders involved, the more important it is to pay attention to forging and sustaining consortia, whether formal or informal, that can act together to further the system's purpose. A consortium provides the various stakeholders, including the intended end users, with a neutral forum in which they can collaborate in developing strategies and approaches to achieve shared goals. This becomes even more important when the various stakeholders have competing interests and different decision-making processes.

10.1.4 System Context: Value of Discovery Engineering

In the systems context (Figure 10.4), the more novel the effort, the greater the likelihood that it will be difficult to predict the behavior of the system with any

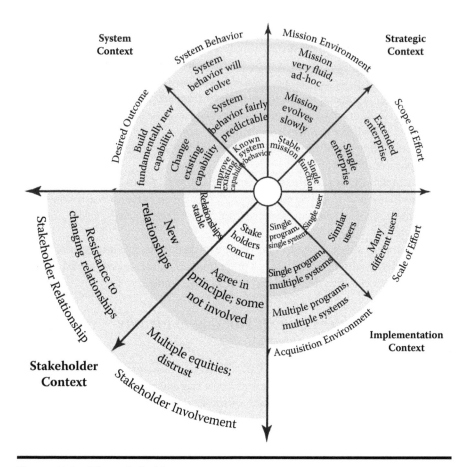

Figure 10.3   The stakeholder context.

degree of confidence until the system is actually developed and deployed. Systems that use technologies that are still maturing are also more vulnerable to unexpected behavior and performance. And the more complex and unpredictable the system behavior, the greater the value of discovery engineering. By discovery engineering, we mean the full range of activities involved in building an understanding of the interactions and behavior of the system of interest. Discovery engineering includes development of prototypes and exploratory integration activities along with early field trials, experiments, and pilots. Prototypes and exploratory integration provide early insight into the technical behavior of the system, while field trials, experiments, and pilots help refine how the system will be used and the interactions between the system and its anticipated users, and allow the systems engineer to assess the impact that the finished system will have on the tasks and processes it supports.

By contrast, the more predictable the system's behavior—that is, the more it lends itself to quantitative assessment and performance modeling—the more

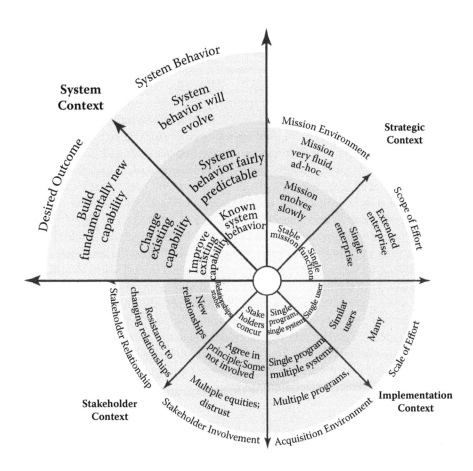

Figure 10.4 The system context.

systems engineers can rely on traditional practices of decomposition, allocation, and integration with a reasonable expectation that the behavior of the components, when they are integrated, will yield the behavior expected of the whole.

10.1.5 Extensions to Current Practice: Managing Uncertainty

The upper hemisphere of the Profiler captures the degree of *predictability* or certainty in the behavior of both the system itself and its operating environment. Traditionally, systems engineering has focused on risk management; that is, it has emphasized identification of the downside risks to program execution and development of contingency mitigation plans to manage those risks. The more uncertain the environment and the more unpredictable the behavior of the system, the more emphasis systems engineers should place on not only managing foreseeable risk, but also exploiting opportunities. Opportunities may arise from evolving user needs and expectations, from changes in the strategic environment, and from the introduction of new technologies. A nascent but growing body of work explores the management of uncertainty (De Neufville, 2004; Hastings and McManus, 2004; Loch et al., 2005, 2006) in terms of implications for system design as well as for project management (see Section 11.5 for a discussion of managing uncertainty).

The lower hemisphere of the Profiler captures *diversity*, in terms of both the number of separate projects that require coordinated activities and the number of different stakeholders and interests involved. Traditional systems engineering integrates and trades off technical and business considerations. Mega-systems engineering must not only continue to balance these competing needs, but must also add *political, operational, and economic factors* into the mix. The greater the number of separate systems and the number of stakeholders with different positions and interests, the greater the importance of including these "soft" issues. Taking these issues into consideration will not necessarily produce better technical solutions. Instead, failure to deal with them increases the likelihood of selecting solutions that will be unacceptable either political or economically, while actively addressing them helps frame objectives and develop feasible approaches that are more likely to gain the necessary support. Further, failure to consider the entire operational context may lead to design solutions that perform well in isolation but fail to perform as part of a larger, more integrated system. In this instance, local optimization does not yield global benefits and may in fact diminish them.

10.2 Matching Practice to Circumstances

This examination of mega-systems has sought to emphasize the need to match practice to circumstances. The Profiler is offered as a tool to help in characterizing these circumstances. It can also be used to reveal the degree of alignment between the particular circumstances and the engineering strategies and practices being followed. Figure 10.5 illustrates a notional example, showing alignment in some segments of the profile (scope of effort, acquisition environment, desired outcome, and system behavior) and misalignment in others (mission environment, scale of effort, stakeholder involvement, and stakeholder relationship).

Failure to match practices to the circumstances at hand can introduce unnecessary friction points in the program's execution. For example, if the mission environment is changing rapidly but the program continues to focus on the documented and approved requirements and fends off changes as "requirements creep," the program risks delivering a capability that may meet contractual *requirements* but has a high probability of failing to satisfy *user expectations*. In such cases, it is not uncommon for the resulting system to be rejected outright or nominally accepted but effectively ignored.

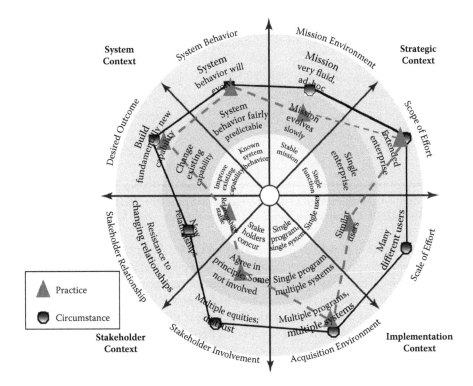

Figure 10.5 Alignment between engineering practice and system circumstances. (From Stevens, R., 2008. *Profiling Complex Systems*, Proceedings of the IEEE International Systems Conference, Montreal, Canada, April 2008. With permission.)

While a high degree of convergence between what the situation demands and the practices being followed does not guarantee a successful outcome, one can hypothesize that significant divergence is an indicator of program misalignment with user needs. As such, the degree of divergence can be viewed as a leading indicator of program health. The greater the total divergence, taking into account both the level of discrepancy within any single wedge and the total number of wedges affected, the greater the magnitude and the breadth of challenges that the program must overcome.

Programs that involve little uncertainty and volatility in user requirements and expectations, and that apply mature technologies whose performance is well understood, are best served if the systems engineer carries out detailed planning and then monitors execution relative to the plan. Programs whose requirements are evolving,

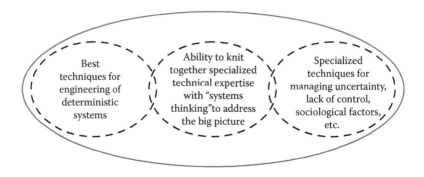

Figure 10.6   A continuum of systems engineering practice. (*Source:* Copyright © 2008, The MITRE Corporation. Used with permission.)

particularly because of changing user environments, would be best served by staged commitment strategies that allow for adjustments to the target and/or the implementation approach. In contrast, programs with an established general vision, but alternative approaches to realizing that vision, may be best served by variation and selection strategies using prototypes, pilots, and other forms of discovery engineering.

A growing body of literature examines the need to match the management and engineering style to the project type. However, considerable further research would be needed to identify the specific techniques and practices that are best suited to different regimes of the Profiler.

10.3 An Emerging View of Systems Engineering as a Continuum of Practice

Figure 10.6 shows an emerging view of systems engineering as a continuum of practice from the techniques best suited to the engineering of purely deterministic systems at the far left to, at the right, the additional techniques needed to engineer complex systems that operate in an uncertain environment. Most cases call for a mix of practices. No set formula can lead to the appropriate mix. Selecting the best approach requires the insight, judgment, and creativity of the systems engineer who understands the system being created, the context in which it is being developed, and the environment in which it will be expected to operate and adapt.

10.4 Refining the Engineering Tenets: A Way Ahead

The emerging tenets outlined below, based on a limited set of case studies, will clearly evolve over time. We offer them as the starting point of a dialogue, not as a definitive body of knowledge. However, they may assist in framing a way ahead.

■ First, *foster the dialogue.* Various organizations and venues can help contribute to and shape the discussions. Universities such as MIT, the Stevens Institute, and the Air Force Institute of Technology have initiated academic programs related to the engineering of large-scale systems and undertaken research to build and refine a body of knowledge. Professional organizations such as INCOSE have encouraged their members to address the emerging challenges of systems engineering in the twenty-first century. Practitioners of systems engineering, whether in for-profit corporations or not-for-profit organizations, have an obligation to examine their practices critically and, where necessary and appropriate, to develop new approaches better suited to this problem space. Customers also have the obligation to demand that practices match the needs of the situation and to question practices that, while well established, do little to achieve their objectives.

Constructive dialogue will emerge through informal exchanges within and among these organizations and their researchers and practitioners. It will be furthered by formal symposia, whether individually or collaboratively sponsored, where the emerging body of knowledge is shared, discussed, and—where necessary—challenged.

■ Second, *agree on a common lexicon.* Today, many terms describe this topic area. We hear "systems of systems," "families of systems," "enterprise systems," "complex systems," and "complex adaptive systems," and we have introduced yet a new term here, "mega-systems." Sometimes these terms are used interchangeably; in other cases, their proponents use different terms to highlight different aspects. Be certain that all participants agree on the meanings of the terms used.

■ Third, *develop a body of case studies.* Case studies provide a repository of individual experiences, lessons learned, and insight into practices that work well—or do not. Case studies are a well-recognized teaching tool in business curricula and are becoming more widely used in engineering education. We need case studies both of successful efforts and—equally if not more important—those that have not succeeded as expected. The value of such case studies will lie not only in the development of a repository of well-documented examples, but also in the potential to discover patterns that provide insight into what works, what does not work, and which circumstances produce which result.

■ Fourth, *refine and extend the engineering tenets.* The principles presented here clearly represent only a starting point and not the expected end state. Dialogue among the larger engineering community, continued research, and lessons learned from a larger body of case studies will certainly help extend this initial set of tenets.

■ Fifth, *define and initiate a research agenda.* Fruitful research opportunities abound. Examples include exploring the implications of complexity science on the development and operations of mega-systems, and identifying and piloting techniques that seem especially well suited to the outer ring of the

Profiler—the messy frontier. Research in the social and behavioral sciences could help in defining more effective approaches to dealing with stakeholder diversity and its impact on the framing of the feasible solution space. Other topics that call for further investigation include the development of flexible and adaptive design tenets and practices, and the transferability of processes and techniques from other fields to systems engineering: For example, would applying some of the principles of adaptive software development to systems engineering processes create a basis for an adaptive systems engineering process that complements the more traditional "V" Model?

■ Sixth, *recognize that practice, not theory, will drive the development of processes and tools.* This is how traditional systems engineering evolved; we anticipate that the engineering of mega-systems will evolve in the same way.

■ Finally, *inculcate a systems thinking mindset.* By that we mean the ability to look simultaneously at the relations and the interactions among the components, the whole system, and the still larger whole in which the system will operate. It means performing trade-offs not only between and among the parts, but also between the parts and the whole (Frank, 2000, 2002).

Ackoff (2004) proposed a definition of systems thinking, which he referred to as "systemic thinking":

> Systemic thinking is holistic versus reductionistic thinking, synthetic versus analytic. Reductionistic and analytic thinking derives properties of the whole from the properties of their parts. Holistic and synthetic thinking derive properties of the parts from properties of the whole that contains them ... when an architect designs a house he first sketches the house as a whole and then puts rooms into it. The principal criterion he employs in evaluating a room is what effect it has on the whole. He is even willing to make a room worse if doing so will make the house better.

As the processes and practices that enable mega-systems engineering emerge and gain acceptance, they will in no way supersede or devalue the practice of traditional systems engineering that has emerged over the past half decade. Instead, they will build on and extend it.

10.5 An Emerging View of Next-Generation Practice

In his book *Rescuing Prometheus*, Hughes (1998) uses the stories of four large-scale projects to illustrate the emergence of systems engineering as a field in the post-World War II era. In his concluding chapter, he presents a table comparing modern and post-modern engineering practices, with post-modern reflecting the introduction of systems engineering. Table 10.1 replicates (with permission) the two columns

Table 10.1   Extending the Comparison: Modern, Post-Modern, and Next-Generation Systems Engineering

Modern[a]	Post-Modern[a]	Next-Generation
Production system	Project	Multiple projects
Hierarchical/vertical	Flat/layered/horizontal	Integrated project teams, consortia
Specialization	Interdisciplinary	Interdisciplinary with information technology as the core
Integration	Coordination	Collaboration
Rational order	Messy complexity	Messier complexity
Standardization	Heterogeneity	Heterogeneity and standardization
Centralized control	Distributed control	Diffused control—limited/no control
Manufacturing firm	Joint venture	Multiple developers / mix of commercial and custom development
Experts	Meritocracy	Community process
Tightly coupled systems	Networked systems	Networked mega-systems
Unchanging	Continuous change	Asynchronous change
Micromanagement	Black-boxing	Standards-based compliance
Seamless web	Network with nodes	Worldwide network
Tightly coupled	Loosely coupled	Mix of tight and loose coupling
Programmed control	Feedback control	Discovery with feedback
Bureaucratic structure	Collegial community	Cross-organization ad hoc collaboration; local interests
Taylorism	Systems engineering	Enterprise systems engineering
Mass production	Batch production	Rapid prototyping, one-off
Maintenance	Construction	Evolution
Incremental	Discontinuous	Spiral
Closed	Open	Open

[a] These columns from Hughes, T.P. 1998. *Rescuing Prometheus: Four Monumental Projects that Changed the Modern World.* New York: Pantheon. With permission.)

from Hughes's work and adds a third column, labeled "Next Generation," that highlights differences and similarities between traditional systems engineering—what Hughes refers to as "post-modern"—and the next-generation practices that have been the focus of this book. It will be interesting to see how closely the actual practices of mega-systems engineering reflect these next-generation characteristics.

10.6 Concluding Thoughts

Mission imperatives, whether in military operations, in homeland security, or in global supply chains, constantly increase the importance and value of information sharing. At the same time, the pervasiveness of information technologies promotes such information sharing, particularly across boundaries. These self-reinforcing factors suggest that mega-systems will become ubiquitous. Whether they succeed or fail in achieving their missions will depend on whether their special characteristics baffle system engineers or prompt them to devise new and creative approaches. We hope that this book will help tip the balance toward the second outcome.

Chapter 11

Postscript: Profiling a Complex Acquisition Program

In the summer of 2006, a government agency invited a study team to apply the Profiler to one of its key acquisition programs. The agency director and deputy director were well aware that their agency faced a number of challenges that were impacting their ability to achieve their mission and execute their acquisition responsibilities within the planned resources and schedule. Moreover, they recognized that their agency was operating in a complex political and economic environment and faced not only these, but also a number of operational and technical challenges. They had been trying to communicate the unique circumstances and constraints facing their program, but believed that they had not been as successful in doing this as they had hoped.

For the agency leadership, the Profiler offered a different approach to articulating and communicating the complex nature of their program to their oversight bodies. For the study team, this study offered an opportunity to test the Profiler in a completely new operational context. The Profiler had initially been developed with an information-centric engineering and acquisition program in mind, but this agency was responsible for designing, constructing, and operating processing facilities at multiple sites in the United States.

The study was a collaborative effort between the agency, its long-time technical advisors, and me (Renee G. Stevens). Together, this team designed a methodology, applied it, and obtained useful results for the agency. The team certainly gained some unexpectedly valuable insights into how the Profiler can be applied to an organization and how it can serve as a diagnostic tool for that organization's activities. Perhaps of greatest interest were the insights about how people in different roles in an organization can view the same situation through different lenses. Neither the agency nor the MITRE team had anticipated these results, but in retrospect they make intuitive sense.

11.1 Multiple Purposes of the Study

The agency had responsibility for several large acquisition efforts. These acquisitions involved complex technologies, and multiple sites operating in various locations, and were affected by a stringent regulatory environment, citizen activists, and an occasionally contentious public debate. In addition, complex federal, state, and local interests play a role in the development and operation of these sites. The program has been plagued by schedule slippages and cost overruns. While there are many reasons for these slippages, many of them beyond the agency's control, the net effect is that the program has had to revise its cost and schedule baseline several times.

The agency leadership fully recognized these challenges and the risks to successful completion of their program. In particular, they were interested in finding answers to the following key questions:

- How well were they managing these risks?
- How well were they communicating the program impacts of these risks to stakeholders?
- Could anything be done either to reduce or effectively manage operational risk?

They saw the Profiler as a tool that could help them delineate the characteristics and risks of their program and communicate these risk issues to the oversight organizations, both in the Executive Branch and in Congress. At the same time, they were interested in identifying internal business practices that could help them better manage these risks.

11.2 Approach

Collectively, the agency and the study team developed an approach in which the Profiler was used as an interview vehicle to capture the perspectives of various managers within the agency and of selected stakeholders external to it. The study team then used the areas of the resulting profile that fell in or abutted the outermost ring as the basis for focusing an assessment of existing business practices.

Table 11.1 Strategic Context Explanatory Language

Mission Environment	
Stable Mission	The environment in which the system is intended to be used does not change over time.
Mission Evolves Slowly	The environment in which the system is intended to be used is expected to change over time, although changes will be relatively modest and infrequent. Some of the changes can be known or anticipated.
Mission Very Fluid, Ad Hoc	The nature of the mission environment changes dramatically and frequently. This may be due to unanticipated mission requirements (new missions, new drivers, and new constraints).
Scope of the Effort	
Single Function	A function is a set of related activities that supports or otherwise accomplishes a mission. Most missions entail multiple functions.
Single Enterprise	The enterprise consists of the entire organization, including all its subsidiaries. The term "enterprise" implies a large corporation or government agency, but may also refer to an organization of any size with many systems and users to manage.
Extended Enterprise	An extended enterprise is a set of independent organizations whose collective efforts are required to accomplish a specific mission.

11.2.1 Tailor the Profiler

The first step was to tailor the Profiler. As originally developed, the Profiler used terminology that is more appropriate to a system development activity, particularly one that is heavily IT based. Despite the considerably different technology, business model, and operational environment, the MITRE team found that it could use the existing terminology with some additional explanation in language that would be understandable to this particular community (see Tables 11.1 through 11.4).

11.2.2 Clarify Sponsor Goals and Objectives

The second step was to meet with the senior leadership of the agency to clarify their goals and objectives. This proved very important in that it not only confirmed their

Table 11.2   Acquisition Context Explanatory Language

Scale of the Effort	
Single User	A single user is an individual or group of individuals who perform similar sets of tasks in the same manner.
Similar Users	Similar users perform a similar set of tasks but do so with some variation. The variation may be a function of local or institutional circumstances.
Many Different Users	The users of the systems perform different sets of tasks.
Acquisition Environment	
Single Program	A single acquisition program that is responsible for the development of a single system type.
Single Program, Multiple Systems	A single acquisition program that results in the development of different systems that *collectively* meet the program's mission needs and requirements.
Multiple Programs, Multiple Systems	A collection of acquisition programs, each of which develops one or more systems that, when integrated, meet a larger program need or set of requirements.

support for the effort, but also highlighted that different senior leaders had different expectations as to how they would use its results. For example, one of the senior leaders was especially interested in using the results as a means of communicating with external stakeholders, particularly those in an oversight role. A second senior leader was more interested in using the results internally. He was most interested in understanding whether the current business practices were aligned with the agency's profile and, if not, what changes would be recommended.

A secondary objective was to determine whether the key managers saw the agency, its mission, and its environment in a consistent manner. In other words, would their profiles be similar or different? This interest in understanding the perspectives of different members of the management team, rather than just the perspective of the senior leadership, led to an approach in which the team used the Profiler as an interview vehicle. Consequently, we found that we needed to develop some type of protocol for conducting the interviews and capturing the results.

Table 11.3 Stakeholder Context Explanatory Language

Stakeholder Involvement	
Stakeholders Concur	Despite having different equities, stakeholders are able to reach agreement on a common set of goals and objectives. Local decisions and actions by individual stakeholder groups are consistent with and supportive of these mutually agreed-to goals and objectives.
Stakeholders Agree in Principle; Some Not Involved	Stakeholders nominally agree to a common set of goals and objectives; however, not all stakeholders can be counted on to act on the agreements. In some cases, stakeholders have not been identified or, if identified, have not been included in the decision-making process. Failure to take into account the equities of all stakeholders may lead to unexpected problems as the system is developed.
Multiple Equities; Distrust	Stakeholders have equities that conflict, and it is difficult to find common ground. Some stakeholders do not believe that their interests will be accommodated. Some stakeholders may be in active opposition.
Stakeholder Relationships	
Relationships Stable	All established relationships among identified stakeholders remain stable over time even though issues may change.
Changing/ New Relationships	Established relationships among identified stakeholders may change in response to existing or new issues. New relationships may be established due to identification of either new stakeholders or new issues.
Resistance to Changing Relationships	Over time, the dynamics between participating groups may require change, increasing the influence of some stakeholders at the "expense" of others. Stakeholders who perceive that their influence is declining will resist these changing relationships, either overtly or more subtly.

11.2.3 Develop Interview Protocol

The protocol that emerged was straightforward. It called for individual interviews with selected managers, senior staff, and key individuals with a program oversight role. The interview itself started with a brief recap of the study purpose and an introduction to the Profiler, followed by a series of questions that stepped the respondent through each of the eight wedges in order. For each wedge, respondents

Table 11.4 System Context Explanatory Language

Desired Outcome	
Improve Existing Capability	*Maintain* an existing capability, but at a lower resource cost or greater efficiency (e.g., new system has higher reliability, thereby decreasing system downtime).
Change Existing Capability	*Improve* an existing capability through adoption of a new technology, application of additional resources, or other means (e.g., increase safety through changes in procedures).
Build Fundamentally New Capability	*Add* a new system capability not present in past designs, typically in response to new requirements.
System Behavior	
System Behavior Known	The operating environment and associated parametric values are established, and the system behavior is known and optimized for these conditions.
System Behavior Fairly Predictable	The operating environment and associated parametric values are generally established and the system behavior is predictable, although not optimized for variability, within these conditions.
System Behavior Will Evolve	The operating environment and associated parametric values may change in response to user response to emerging needs that may not be well defined. System behavior (design) will evolve in response to changing user needs.

were asked to mark a blank Profiler and indicate the spot that best corresponded to their assessment of the current situation. Some respondents chose to center their answer in the ring that corresponded to their assessment. Others took the opportunity to provide more nuanced responses by placing their assessments on or near the border between adjacent bands. As interviewees discussed the rationale for their assessment, the team captured the information using a tape recorder as well as one or more note takers.

We also asked the respondents to indicate whether they anticipated any changes over the next 3 to 5 years and the direction of such change, either outward or inward. Interestingly, some respondents identified more than one factor that might influence the situation, with one factor causing an outward move toward greater complexity and uncertainty and another factor that might cause inward movement.

In parallel with developing the protocol and supporting material, the program office identified the list of individuals they wanted us to interview. The initial set of interviews was limited to managers at different positions inside the agency. At the direction of senior management, the list was subsequently expanded to include staff in higher headquarters and on Congressional committees with an oversight role for this program. While these staff could be considered stakeholders, this list was clearly limited and did not include many external stakeholders.

11.2.4 Conduct the Interviews and Synthesize the Responses

Before starting the interviews, senior agency leaders scheduled a kick-off meeting with their key staff to familiarize them with the objectives of the study as well as the key elements of the Profiler. Senior leadership participation and their obvious interest and support proved critical in scheduling and conducting the interviews and in gaining the cooperation of the individuals selected as interviewees.

Where possible, interviews were conducted in person and were held separately with each manager being interviewed. Where that was not possible, either because of travel requirements or schedules, the team did accommodate telephone interviews. In all cases, one person asked the questions, and one or more others took notes. Interviews were scheduled to last 90 minutes and typically used the full time allotted. After the interviews, the study team consolidated the notes and sent them back to the interviewee for confirmation and, where necessary, clarification.

After trying different approaches to synthesizing the individual responses, the study team settled on an approach that organized the responses by the respondent's role within the organization. Four different views were developed:

1. Site view, as provided by managers who were responsible for operation of a particular geographic site
2. Project view, as provided by managers responsible for the design, development, and acquisition activities associated with a particular technology or project
3. Agency view, as provided by senior managers and those with an agency-wide function
4. External view, as provided by senior staff with a program oversight role at the department level as well as with cognizant Congressional staff (because of schedule conflicts, only one of the identified Congressional staff members was available to be interviewed)

11.2.5 Conduct Workshops with Key Agency Staff and Leadership

In addition to the introductory session, the team held two workshops: one about midway through the study and the second one near the end of the study. These workshops were initially intended to serve as in-process reviews, providing a status update to agency staff and leadership. The workshops proved particularly useful. Not only did they provide an opportunity to present emerging findings, but they also served as a forum to highlight potential anomalies in the profile responses that warranted reexamination, to identify additional individuals who should be interviewed, and to reach agreement on those areas of the profile that called for a more in-depth review and assessment of current business practices.

11.2.6 Review and Assess Agency Business Practices

Because one of the key purposes of the study was to identify business practices that could be used in reducing or effectively managing operational risk, the team focused its efforts on those parts of the agency profile that fell in or near the border of the outermost ring because these were the areas that respondents had highlighted as involving the greatest complexity and uncertainty. The team identified, researched, and documented business practices that could be instituted to address the targeted focus area and, in parallel, researched the agency's existing business practices. Where possible, the preferred approach was to leverage and augment existing business practices.

11.3 Findings

Organizing the interview responses by role within the agency provided a useful way to structure the findings. It also helped in identifying some potentially significant response patterns that highlight how managers and staff within the same organization read the situation through the lens of their particular roles and responsibilities. In general, the team observed in this study that the profiles generated by respondents with broader responsibilities tended to be more expansive; that is, they had more points in or abutting the outer ring, while those with more focused responsibilities tended to be more compact, with more points in the innermost and middle rings.

11.3.1 Role-Specific Profiles

The following subsections step through the profiles for each of the management roles. Then we discuss the similarities and differences between these separate profiles and suggest some reasons for these differences.

11.3.1.1 Site Manager Profile

The site view represents the perspective of managers who are responsible for running operations at a particular geographic site. They deal primarily with technical and operational anomalies that impact their ability to achieve and maintain planned throughput rates. In general, they view these as "manageable" risks, in

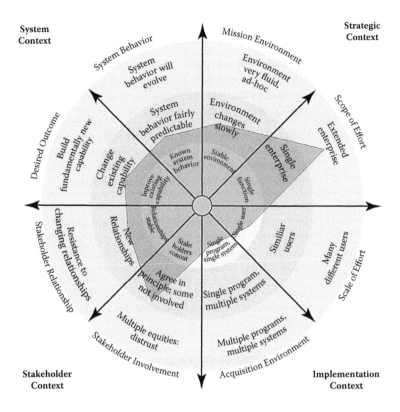

Figure 11.1   Site manager profile.

that they have the flexibility to adapt procedures to deal with them. On the other hand, issues related to interactions with organizations outside the agency's direct control (extended enterprise) and with stakeholders are seen as areas over which site managers have less control. Figure 11.1 shows the site manager's composite profile. Note that this profile is fairly compact, with only two points that are near or in the outer band (extended enterprise and stakeholder involvement). The factors underlying this profile are summarized in Table 11.5.

11.3.1.2 Project Manager Profile

The project manager profile is based on information collected from a diverse group of project managers, deputy project managers, and business managers from different areas of the agency. Their composite profile (see Figure 11.2) is

Table 11.5 Site Manager View

Mission Environment	The mission environment changes in response to both internal and external events. Internal events include unanticipated anomalies during processing that may result in delays. External events are typically changes to governing regulations that require reengineering some equipment and modifying some operations.
Scope of the Effort	The program was viewed as an extended enterprise in that, to be successful, it required the cooperative efforts of organizations at the federal, state, and local levels that fell outside the authority and control of the agency.
Scale of the Effort	Site managers interpreted the term "user" to mean plant operators who were expected to follow standard operating procedures. However, there were enough differences among sites that suggested that there was, in fact, a community of similar users rather than a single class of users.
Acquisition Environment	From the site perspective, these managers viewed the acquisition effort as a single program producing a single system.
Stakeholder Involvement	Site managers work very hard to develop and maintain good relationships with local communities. While specific issues may perturb these relationships temporarily, they tend to return to a steady state where stakeholders agree in principle. Relationships with local stakeholders have improved considerably, and that trend is projected to continue.
Stakeholder Relationships	Stakeholder relationships are considered stable but are subject to change based on the specific issue at hand.
Desired Outcome	Site managers see the desired outcome as improving the existing capability with the goal of increasing throughput.
System Behavior	There are two technologies being used. One is considered a mature technology, and the system behavior is well understood. The second technology is viewed as fairly predictable, and the focus is on achieving a higher level of stability in system behavior.

more expansive than that of the site manager, with most of the points falling in the middle band. Like site managers, project managers focused on technical and operational risk and also addressed political risk. Again like the site managers, project managers placed Scope of the Effort and Stakeholder Involvement in the outermost band and added a third one, Desired Outcome. Unlike site managers,

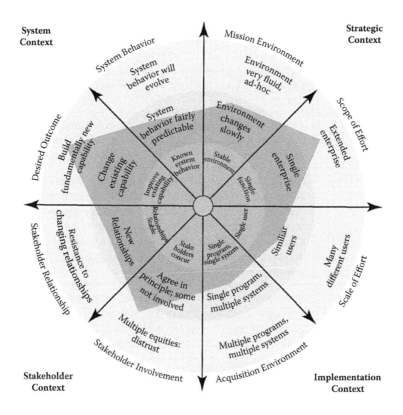

Figure 11.2 Project manager profile.

who focused on improving or changing existing capability, project managers tended to see their role as being centered on technology innovation. Table 11.6 summarizes the responses.

11.3.1.3 Agency-Wide Profile

The agency profile was constructed from interviews with headquarters staff who had program-wide or cross-cutting responsibilities. Interviews conducted with the agency director and other members of the senior leadership were not included in this composite. Like that of project managers, the agency-wide profile fell in the middle and outermost rings with three areas entering the outermost ring (Scope of the Effort, Stakeholder Involvement, and Desired Outcome). However, it is a little

Table 11.6   Project Manager View

Mission Environment	Most respondents characterize the mission environment as evolving slowly. They also acknowledge that particular technical or operational issues affect the mission environment, particularly when the environment is politically charged.
Scope of the Effort	All but one of the respondents saw this agency as operating in an extended environment. They did not foresee changes to this situation.
Scale of the Effort	Like site managers, project managers view the plant operators as the users. Respondents who came from projects with a single technology focus and with common design and operating procedures tended to identify a single user group. Respondents who fielded different technologies or recognized that their plants were configured or operated differently identified the population as "similar."
Acquisition Environment	Responses differed widely, depending on whether a project manager chose to address this question from a project-specific perspective or from a larger mission or even agency-wide perspective. In some instances, respondents viewed each individual site within the same program as a separate system. Those taking the broadest perspective viewed the agency as operating in an acquisition environment of separately managed programs that are developing multiple systems.
Stakeholder Involvement	Project managers characterized stakeholder involvement almost evenly between the middle (Agree in Principle) and the outermost (Multiple Equities, Distrust) bands. One respondent noted that stakeholders had a wide range of interests, not all of which were supportive of the program or consistent with its needs. Factors that influenced stakeholder views of the program include general mistrust of government as well as concerns about specific local environmental and economic issues. While activists were viewed as a source of distrust, project managers also expressed the opinion that relations with local communities were improving over time.
Stakeholder Relationships	Like site managers, project managers considered "Stakeholder Relationships" stable but subject to change based on the specific issue at hand. This situation is not expected to change, as it is likely that new issues will surface even as other issues are addressed and resolved.

Table 11.6 (continued)   Project Manager View

Desired Outcome	Unlike site managers, who see the desired outcome as improving the existing capability, project managers generally characterize their goal as building a fundamentally new capability. This apparent discrepancy reflected the project managers' focus on the development cycle, with particular attention to the development or application of new technologies.
System Behavior	Respondents generally reported that system behavior, independent of the specific technologies being used, was fairly predictable. They noted that system behavior was predictable by design, and that sufficient margin was built in to allow for variability. More than half of the respondents saw an inward trend toward increasing predictability. One in particular noted that increased stability was gained by operational experience and application of operational lessons learned.

more expansive than the project manager profile (see Figure 11.3). Table 11.7 summarizes the results.

11.3.1.4 External Stakeholder Profile

The external stakeholder profile[1] captures the responses of staff members from external offices that have program management or oversight responsibility over this agency. It is, by far, the most expansive of the four profiles, with more responses being in or near the border of the outermost ring (see Figure 11.4). These respondents acknowledged the technical and operational uncertainties associated with some of the technologies, but focused their remarks on their political and economic consequences. In effect, they viewed this as a technical program that is heavily influenced by political considerations at the local, state, and federal levels. Table 11.8 summarizes the responses.

11.3.2 Similarities and Differences in the Agency Profiles

When comparing the role-based profiles, the study team saw both striking similarities and differences based on respondents' roles and responsibilities (see Figure 11.5). Independent of role and responsibility, most respondents had a common perception of key areas of complexity and uncertainty facing the agency.

The agency is consistently viewed as operating in the outermost ring in two areas. In the Strategic Context, it is described as an extended enterprise in that its

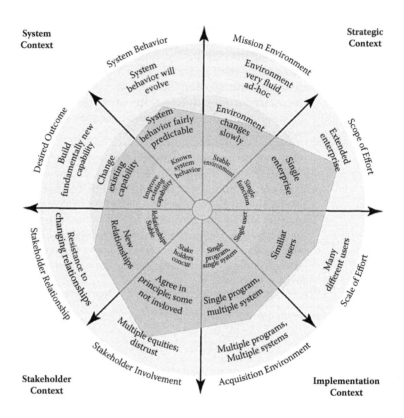

Figure 11.3 **Agency manager profile.**

operations and ability to remain on budget and schedule depend on external partner agencies over which it has no control and little direct influence. Respondents at all levels were able to identify the same set of external agencies. In the Stakeholder Context, the agency is viewed as having to deal with multiple stakeholders who have different and often opposing interests, and who distrust each other's motives and actions. While respondents noted that the situation had improved over the initial baseline, they nevertheless continued to view this dimension as one that continues to perturb agency

> There is a common perception of key areas of complexity and uncertainty facing the agency.

Table 11.7   Agency Manager View

Mission Environment	Respondents generally characterized the environment as changing slowly over time. One noted that the environment was generally stable except for periods in which particular events threw it into the very fluid region. Once these precipitating events were resolved, the environment reverted to a stable state. Of interest, agency respondents saw the mission environment as having technical, organizational, and regulatory dimensions.
Scope of the Effort	Respondents consistently viewed the agency as operating in the extended enterprise or on the border between that and a single enterprise. All respondents were able to identify agencies that provided support to or exercised some degree of influence over the agency and how it conducted its mission.
Scale of the Effort	Most respondents viewed users as similar but differentiated based on site-specific technologies and business practices.
Acquisition Environment	As in the case of project managers, respondents viewed the acquisition environment differently, depending on whether they were addressing their particular area of responsibility or that of the larger agency.
Stakeholder Involvement	All of the respondents recognized the range of stakeholders as including local communities, citizen activists, regulatory bodies, other federal agencies, and the agency's higher headquarters. Those who characterized this area in the outermost band (Multiple Equities, Distrust) primarily cited the role of citizen activists who, while a small minority of stakeholders, exerted a considerable impact on agency strategies, plans, and activities.
Stakeholder Relationships	Stakeholder relationships were characterized as bordering on or falling in the middle band (New Relationships). State regulators and local communities opposed to the program had formed alliances. However, as both had gained confidence in site operations, opposition to the program eased and these alliances became less prevalent. At the same time, local business communities have become supporters as they recognized the economic benefits.

(continued on next page)

Table 11.7 (continued)   Agency Manager View

Desired Outcome	In general, respondents saw the agency as focusing on changing an existing capability. This is perhaps best exemplified by one respondent who described the effort as a continuous state of process improvement but not one of major technological change. Others saw the program as having pioneered fundamental changes in technology that have since become well established in the industry.
System Behavior	Most respondents characterized this program as being in or at the border of the fairly predictable region. Specific factors, including unanticipated events that resulted in operational delays, changes in environmental regulations, and unknown consequences of site closures, were cited as factors that precluded future behavior from being completely known.

operations. By contrast, areas that were in or near the middle ring were generally ones that respondents believed might entail risks, but ones that they could manage either technically, programmatically, or operationally.

At the same time, the team saw some interesting patterns in the profiles associated with the different categories of respondents. Site managers consistently had the most compact profiles, with more points in the innermost and middle rings. Project managers and agency-wide managers tended to have successively more expansive profiles. In effect, the broader the role, the more likely the respondent was to view the program as being more complex and more subject to external influences. This pattern is supported by the observation that site managers were more likely to emphasize technical and operational factors that interrupted plant functions or otherwise impacted their ability to stay on schedule. They viewed these as disruptive but manageable. Project and agency managers also addressed technical and operational issues, but tended to add political and economic factors as further considerations. In contrast, external respondents tended to view issues almost exclusively through political and economic lenses. In fact, technical and operational issues were often viewed and judged in terms of their political and economic consequences. Such political and economic factors can, and do, constrain the technical and programmatic options that are available to the agency. Options that may be technically feasible and cost effective may not prove viable if they encounter community or political opposition. Similarly, local interests may drive the program to implement technical options that entail significant cost and schedule consequences.

11.4 The Profile as an "Uncertainty Map"

As respondents discussed areas they considered as being under their control and manageable, and contrasted them with other areas over which they had little

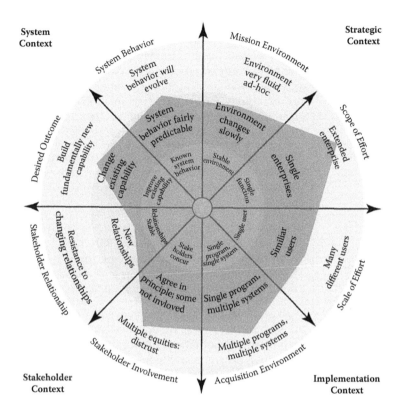

Figure 11.4   External oversight profile.

control, the notion emerged of the profile as an uncertainty map. Respondents who characterized the situation as being closer to the center tended to focus on those issues or events that were primarily internal to the program and agency. Examples included contaminants introduced during processing and lapses in following standard operating procedures. These events may have interrupted operations, resulting in schedule slippages and increased costs; but once recognized, the problems could be addressed by technical measures or by changes

Respondents with broader roles tended to have broader profiles. They were more likely to see the program as being more complex, more uncertain, and subject to external influences over which they had little control.

Table 11.8   External Agency View

Mission Environment	Respondents typically indicated that while the mission of the agency is clear, the environments in which it is conducted tended to border on the fluid, primarily because of the introduction of new technologies and political considerations. One respondent noted that the mission environment changes slowly and was trending inward toward greater stability.
Scope of the Effort	All respondents viewed the program as operating in the Extended Enterprise realm because external organizations, both partners and stakeholders, had an effect on its operations.
Scale of the Effort	Respondents generally believed that users were performing the same function but, because of differences in technologies or site-specific variations, they were performing it differently. All respondents indicated that they anticipated a convergence toward similar users but that this would take some time.
Acquisition Environment	Most of the respondents saw the overall program as falling in the Multiple Programs, Multiple Systems region, acknowledging both the existence of two separate acquisition programs within the agency as well as the different technologies that were in operation or under development.
Stakeholder Involvement	All the respondents saw Stakeholder Involvement as falling in the outer band and identified the same stakeholder groups, including both internal and external groups. External groups included federal and state regulators and interest groups. One of the respondents saw the stakeholder involvement trending inward over time while others anticipated no change.
Stakeholder Relationships	Several respondents indicated that most stakeholders have well-established relationships that do not change with the issues. However, they did acknowledge that new stakeholders might emerge as the program moves through its life cycle. Another stakeholder noted that community outreach efforts have been successful, and that there is a growing tendency toward local support for the program as initial fears are eased.

Table 11.8 (continued) External Agency View

Desired Outcome	A wide variety of opinions were provided, ranging from continuous improvements in existing operations (improving existing capability), to the introduction of new technologies (changing existing capability) and, for the program as a whole, to providing a fundamentally new capability.
System Behavior	Most respondents indicated that system behavior was fairly predictable, although operations must constantly adapt in response to unanticipated events, both internal and external. In general, they saw the situation improving over time and becoming more predictable.

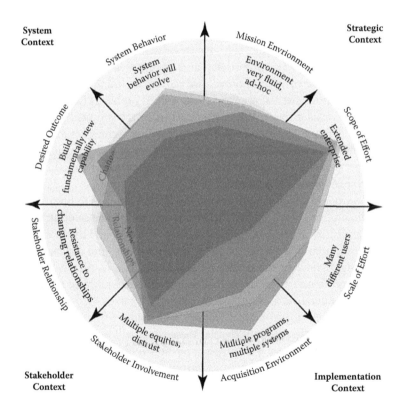

Figure 11.5 Composite views.

in operational procedures that these respondents saw as falling within their area of responsibility. In these cases, the managers were in a position to devise and apply specific mitigating strategies. Once dealt with, the issue was unlikely to recur, although it was possible—and indeed expected—that new problems might surface. In general, respondents reported that increased experience in operations create a trend toward greater stability and fewer unanticipated interruptions.

In contrast, respondents who characterized the situation as being closer to the outer ring tended to focus on issues that were derived from the external environment, such as dependence on external partner agencies or stakeholder interaction, or issues that dealt with highly uncertain future events. These "Messy Frontier" aspects of the situation generated higher levels of system instability, and did not typically lend themselves to management by existing practices. Most often they concerned political and economic factors that these respondents saw as falling outside their ability to control. Thus, a key finding of this study was that the external environment had a critical influence on this agency's ability to accomplish its mission.

Systems engineers should consider this distinction between uncertainty that is primarily technical and operational and therefore more manageable and uncertainty that is political, organizational, and economic, and therefore less manageable, parallels the earlier discussion of tame and wicked problems (see Section 4.3). They should also consider that different participants in the process have quite different perspectives around issues and the most appropriate techniques to deal with them. The findings here suggest that these differences are not just personality based, but may in fact be closely aligned to the particular role the individual has in the project's engineering and management.

Technical and operational issues usually lend themselves to linear approaches in which the problem is recognized, criteria for an acceptable solution are identified, facts are collected, alternative approaches are postulated, and the assessment yields a workable solution. By contrast, political, organizational, and economic issues, such as those highlighted in this study, are less amendable to such a linear approach. By their very nature, approaches to such uncertainties must be more broadly collaborative and iterative. Given the different and (in some cases) competing interests involved, different organizational equities and modes of interaction and different definitions of success, it is unlikely that these issues will lend themselves to a single-pass solution. Rather, they will reappear, possibly under somewhat different guises.

In viewing the profile as an "uncertainty map," we do not mean to imply that there is no uncertainty or no risk in projects whose profiles are closer to the center ring. Instead, the message is that the nature of uncertainty is fundamentally different in different portions of the Profiler. Near the center, it is more manageable and lends itself to well-developed engineering and management approaches. As one moves outward, not only does the nature of the uncertainty change, becoming more unpredictable and less amenable to traditional methods, but the approaches to dealing with it must also necessarily change.

11.5 Recommended Practices for Dealing with Uncertainty

We focused our review and assessment on business practices that could be used to deal with uncertainty, particularly uncertainty that is externally driven and over which the agency has little direct control. We focused on three interrelated practices: uncertainty management, stakeholder mapping, and environmental scanning.

11.5.1 Characterizing Uncertainty

Hastings and McManus (2004) define uncertainties as "things that are not known, or known only imprecisely."

There are many sources of uncertainty. They can be internal as well as external.[2] Even in predictable, well-understood projects, internal uncertainty can derive from the specifics of the work to be done or the technologies to be implemented. It can encompass technical risk as well as unpredictable or unpredicted interactions among the components of the system. Novel projects or projects using new and emerging technologies inherently carry more uncertainty than those that rest on a solid foundation of experience. Novel projects are based on a set of assumptions—often on far more assumptions than actual knowledge (McGrath and MacMillan, 1995). Those assumptions may prove unfounded and conditions may alter substantially from the initial situation. Similarly, technologies may change and requirements may evolve.

Uncertainty can also derive from fluctuations in the external environment. These could include changes in the operational context, user needs, governing policies, regulations, or markets, as well as the actions and influence of external stakeholders. Project leaders have considerably more control over internal uncertainties in that they can be more readily resolved with resources that are at the organization's disposal. Externally based uncertainties are, by their very nature, difficult to control.

One can view uncertainty along a continuum ranging from incomplete knowledge to complete lack of knowledge (Figure 11.6):

- *Incomplete knowledge.* Knowledge to be supplemented by other information that is available but has not yet been collected or determined.
- *Variability.* Factors that can have a range of potential outcomes, such as the duration of particular activities. Variability is often associated with project parameters such as cost, schedule, and performance.
- *Foreseeable events (known unknowns).* These are events that can be anticipated, although the likelihood of occurrence is unknown. Foreseeable events are often handled via preplanned alternative paths triggered by the event's occurrence.

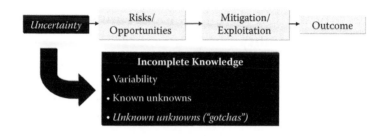

Figure 11.6 Uncertainties, risks, and opportunities.

■ *Unforeseeable events (unknown unknowns).* These are the events or interactions that are truly unpredictable and therefore not anticipated or accounted for. They are often found in novel projects or in circumstances where the external environment experiences radical and unexpected shifts. Hastings and McManus (2004) refer to these as "gotchas."

Uncertainty, in and of itself, is neutral. Uncertainties can produce risks, which are generally viewed negatively as threats to successful execution of the projects or as problems to be mitigated. However, they can also create opportunities, which are generally viewed positively as situations to be exploited (White, 2006a, b).

While project management organizations such as the U.S. Project Management Institute (PMI) and the United Kingdom's Association for Project Management (APM) generally adopt a neutral definition of the term "risk," encompassing both downside and upside effects (PMI, 2001), a tendency remains to think of risk primarily in terms of negative consequences that call for mitigation. The *Risk Management Guide for DoD Acquisition* (DoD, 2003) defines risk as a "measure of future uncertainties in achieving program performance goals and objectives within defined cost, schedule, and performance constraints." Table 11.9 highlights how different kinds of risks can, under different circumstances, turn out to be opportunities.

11.5.2 Techniques to Deal with Uncertainty

This section discusses three techniques for dealing with uncertainty, particularly uncertainty that derives from external events and external stakeholders:

1. *Environmental scanning:* Techniques to identify changing trends and patterns in the external environment and their implications for the organization. Implications can be either positive or negative (or in some instances, both).
2. *Stakeholder analysis and management:* Techniques to identify people, groups, and organizations that can influence actions and outcomes, assess their interests, and formulate appropriate forms of engagement.
3. *Uncertainty management:* Extension of well-established project risk management practice to manage both risks and opportunities at the project and enterprise levels.

Table 11.9   Risks and Opportunities

Risks	Opportunities
Disaster: System causes harm.	Not applicable.
Failure: System does not work.	*Emergent capabilities:* System works exceptionally well and/or for purposes not originally envisioned.
Degradation: System works but not up to initial expectations.	*Unexpected capability:* System exceeds expectations.
Deviations: Program runs over cost or schedule.	Programs are almost never early or under budget.
Market shifts: System works but the need for its services is less than assumed level.	*Market shifts:* System works, and the need for its services is more than assumed level.
Need shifts: System works but does not meet new needs.	*Need shifts:* Need for system increases and/or new needs are uncovered.

Source: Adapted from Hastings, D., and H. McManus. 2004. A Framework for Understanding Uncertainty and its Mitigation and Exploitation in Complex Systems. Presented at *MIT Engineering Systems Symposium.* Cambridge, MA. http://ocw.mit.edu/NR/rdonlyres/Engineering-Systems-Division/ESD-85JFall-2005/695980D5-FDEE-4C4D-A505-14B8E9C95841/0/uncertainty.pdf (accessed February 10, 2008).

Figure 11.7 depicts the relationship between these three techniques. The first two techniques—environmental scanning and stakeholder analysis and management—help identify the nature and extent of external events and players, thereby reducing, although clearly not eliminating, the element of surprise. Uncertainty management builds a portfolio of actions to address these uncertainties.

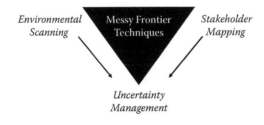

Figure 11.7   Techniques to deal with uncertainties.

11.5.2.1 Environmental Scanning

Environmental scanning is fundamentally about gathering and using information about events, trends, and relationships in an organization's external environment. In effect, it provides managers and executives with a level of situation awareness not only about the current environment, but also about how that environment may change in the future. Environmental scanning is used to identify potential threats and opportunities that may affect the organization, its strategy, and its mode of operation. Most typically, it serves as a key input to strategic planning at the enterprise level.

> Environmental scanning provides situational awareness of the external environment.

While the range of topics that can be included in an environmental scan is quite broad and will necessarily be tailored to the particular organization, its circumstances, and its market, in general they can cluster into four broad categories: Political, Operational, Economic, and Technical. (Note that these four categories form the acronym POET.) Examples of environmental scanning topics within each of these four categories include the following:

1. Political and cultural (including statutory, regulatory, and policy changes, as well as changes in community attitudes toward the organization)
2. Operational (including emerging drivers and constraints of current and future operations)
3. Economic (including macro-economic and/or local economic conditions, relevant demographic changes, labor market changes, and changing budget pressures)
4. Technical (including emergence of new technologies or obsolescence of existing ones and shifts in productivity and infrastructure)

Environmental scanning can be conducted on an ad hoc basis, periodically to support a particular planning activity, or as a continuous process. Ad hoc or periodic scans make the most sense when the external environment is stable or changes relatively slowly. Continuous scans are most useful in those situations in which the environment is highly fluid and subject to significant shifts. Most organizations link environmental scans to their strategic planning calendar.

While there is no specific template or agreed-to process for conducting environmental scans, the following are some implementation considerations:

■ Ensure that there is executive-level support for the scan.
■ Assign a focal point.
■ Establish a small team to conduct the scan.
■ Select team members on the basis of their "big-picture" perspective, willingness to look beyond known terrain, and ability to spot early warning signs.

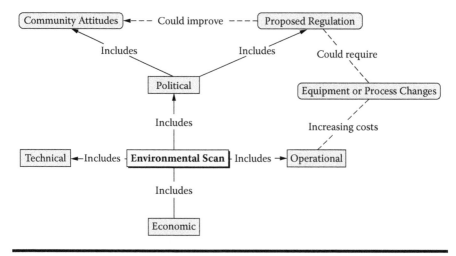

Figure 11.8 Example concept map showing external events and their possible impacts.

- Balance the size of the team with heterogeneity. Heterogeneity is an asset to be leveraged because it fosters different perspectives.
- Decide what topics to include, but do not eliminate topics too early. Be prepared to add topics that were not initially considered.
- Collect data: read, talk to knowledgeable people, and go beyond normal contacts.
- Prepare brief abstracts highlighting the trend and its possible implications for the organization. Note that the same event or trend could have both positive and negative impacts.
- Use concept maps or other techniques to show the trends and the possible relationships across trends; see example in Figure 11.8.
- Present the emerging findings to the executive team. Incorporate their insights and perspectives.
- Incorporate results of the environmental scan into the organization's strategic risk and opportunity management process.

11.5.2.2 Stakeholder Analysis and Management

Originally developed by the business community, stakeholder analysis and management constitute a straightforward methodology that seeks to account for the interests of stakeholders in the achievement of a desired outcome, whether that outcome is the development of a policy or the successful completion of a project. Stakeholder analysis forms the basis of action plans that are geared toward increasing support for a proposed course of action or, where necessary, toward reducing opposition to it. While stakeholder analysis is most often conducted at the start of a project, changes in the circumstances, the issues, the players, or their interactions may warrant a periodic reexamination.

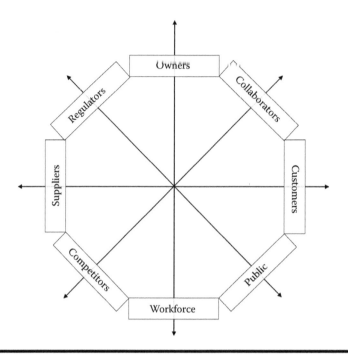

Figure 11.9   Eight types of stakeholders.

Stakeholders are individuals, groups, or organizations that have a legitimate interest in a project or organization. Stakeholders can be internal as well as external and include the following:

- Those whose interests are affected by the project or entity or whose activities affect it
- Those who have information, resources, and expertise needed by the project
- Those who control relevant implementation instruments, such as funding or legal authority

Stakeholders can also be viewed in terms of their roles (see Figure 11.9)[3]:

- *Owners (or leaders)* seek to achieve the vision and mission of the organization. They may be the executive team inside the organization, or the individuals and organizations that provide external oversight.
- *Collaborators* have common or intersecting interests but are independent of the organization. Their actions are critical to achieving the organization's goals.
- *Customers* are those that ultimately use the goods and services provided by the organization.
- The *public* in general, and the local community in particular, are affected by the actions of the organization and have a set of sometimes conflicting attitudes to and interests in these actions. For example, a community may

be concerned about the environmental impacts of an operation while, at the same time, welcoming the economic benefits that the organization brings to their community. Included in this stakeholder group are local citizens' groups, public interest groups, and activists.

■ The *workforce* includes the employed workforce as well as the potential labor pool. Workers are interested in maintaining the economic benefits of employment while also being concerned about stability and safe operations, among other factors.

■ *Competitors* seek to substitute their products and ideas for those of the organization. In a typical market situation, they would seek to appropriate the organization's market, customers, employees, and proprietary information.

■ *Suppliers*, including contractors, seek to gain maximum return for the goods and services they provide while maintaining or extending their market.

■ *Regulators* seek to achieve compliance with laws or policies. While regulators primarily come from government—federal, state, and local—they can also include nongovernmental organizations with policy and standards authority such as industry, trade associations, and voluntary standards consortia.

Stakeholder analysis entails four key steps:

■ *Stakeholder identification* involves listing all the potential individuals, groups, and organizations that are affected by, have an interest in, or can influence the organization's operations.

■ *Stakeholder prioritization* helps to identify the key stakeholders, sorting them according to their importance. While there are a number of templates that help management visualize relative stakeholder importance, perhaps the most commonly used one is a simple power-interest grid (see Figure 11.10). High-priority stakeholders are those with both high power and high interest. They are the ones who warrant especially close management. Stakeholders with high power and low interest as well as those with high interest but low power should be kept informed, while those with low power and low interest require minimum effort. Note that this kind of mapping can be done for the whole project or particular parts of it.

■ *Stakeholder analysis* defines the stakeholder's position relative to the organization or issue at hand. This position can be understood as being located along a continuum from strong opposition to strong support. Stakeholder positions can be depicted by color-coding on the stakeholder map. Analysis is also used to identify the factors that motivate stakeholders, both individuals and groups, to determine who influences them as well as who they, in turn, influence.

■ The *stakeholder management plan* identifies the specific actions necessary to convert stakeholders from critics to supporters (or at least neutral participants), manage the opposition from those who are expected to remain critics, and retain supporters.

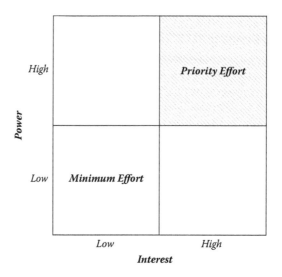

Figure 11.10 Power interest grid.

11.5.2.3 Risk Management and Uncertainty Management

Risk management is an integral part of project management. It has developed into a distinct discipline with formalized, well-documented techniques, procedures, and tools,[4] as well as associated professional bodies, guidelines, and training materials. It is also recognized as one of the key technical management practices of systems engineering. It is one of the nine knowledge areas in the *Program Management Book of Knowledge (PMBOK)*, and its application to the DoD is documented in the *Risk Management Guide for Department of Defense Acquisition* (DoD, 2006).

While risk management most often focuses at the project level (and is consequently termed "Project Risk Management"), some risk management practitioners and researchers express a growing interest in extending the practice across large organizations. Terms such as "strategic risk management" and "enterprise risk management" are being introduced. At the same time, there is also interest in extending the discipline to address explicitly both the consequences with negative impacts and those with potentially positive ones. These topics have generated considerable debate within the risk management community, much of which is being captured in various presentations and publications. It is, however, important to emphasize that this is still an emerging practice area, and there is no widespread agreement on terminology, let alone on the practices to be implemented.

Figure 11.11 captures the dimensions of this emerging risk management landscape. Starting at the lower left, the figure highlights three paths that are being explored. Path 1 is the extension of Project Risk Management beyond the project level to the organizational or even the enterprise level. In general, these approaches

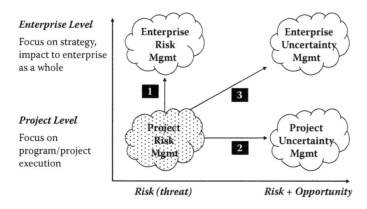

Figure 11.11 Emerging paths in the management of risk and uncertainty.

continue the fundamentals of risk management processes, techniques, and tools and seek to apply them to risks that may impact the organization's strategy. While it extends the focus of the activity, it does not generally change definitions. Path 2 is the expansion of Project Risk Management to include both risks and opportunities and to add unforeseeable events (unknown unknowns) to the range of uncertainties. Although there is no agreement on terminology, some authors use the term "Uncertainty Management" to refer to the management of risks and opportunities.

Path 3 effectively combines Paths 1 and 2. It seeks both to extend the discipline to organizational levels above the project and to expand the focus from an emphasis on downside risk to one that encompasses both risks and opportunities. To some extent, uncertainty management at the enterprise level is closely related to strategic planning, particularly strategic planning under uncertainty (*Harvard Business Review* staff, 1999).

These trends suggest the need for a broader view of risk management that encompasses the following:

- *Opportunities as well as risks.* Opportunities may arise from new technologies that increase the effectiveness or efficiency of operations, new missions, or new operational practices. Opportunities may also arise from changes in the external context, including changes in stakeholder positions or relationships.
- *Unforeseen events (unknown unknowns) as well as foreseen events.* Foreseen events are the ones that we can envision and for which we can develop contingency plans. Unforeseeable events are, by definition, unexpected. They can occur during any phase of the program (planning, design, operations, or closure). They can be internal, such as unexpected operational challenges, or they can come from external events and factors, such as unexpected shifts in political support.

■ *Strategic perspective (top down) as well as project execution perspective (bottom up).* Many of the risks to the execution of the program as planned are best understood at the level responsible for implementation. At the same time, many of the risks and opportunities that affect the overall enterprise are often best understood at the executive and senior manager levels of the organization. These risks and opportunities are strategic in nature, in that they are broad based, affect the achievement of the enterprise's essential goals, and may require fundamental changes in strategies, plans, or approaches. An approach that combines a bottom-up perspective to provide the necessary implementation and operational realism and a top-down perspective to provide the more encompassing strategic view offers a more comprehensive framework in which to identify and manage uncertainties.

11.5.3 Putting Ideas into Practice

Uncertainty management is not merely an extension of existing practices. To a large extent, it demands a fundamentally different mindset. In effect, it entails "expecting the unexpected" (Loch et al., 2006) and being prepared to adapt to it. It means looking for early indicators of changing conditions, assessing their implications for the project and the organization as a whole, determining the set of actions to be taken to address these uncertainties, and actively monitoring both the situation and the effectiveness of strategies and actions in dealing with it. Figure 11.12, based on the work of Courtney et al. (1995), summarizes this approach to confronting uncertainty.

Figure 11.12   Approach to confronting uncertainties.

The study conducted with the Profiler recommended several specific practices to deal with the nature and range of uncertainties confronting this particular agency. Specific recommendations included the following:

- Establish an executive-level team to focus on strategic risks and opportunities. Meet regularly to review the situation and adjust priorities.
- Establish an environmental scanning process to look for early indicators of changing external conditions.
- Establish a stakeholder analysis and management process to identify key stakeholders, their positions, and specific actions to be taken to improve support and/or mitigate resistance.
- Focus stakeholder efforts on key stakeholders. Seek out opportunities to maintain direct communications not only with key supporters but also with critics.
- Track the effectiveness of actions taken to mitigate risk and exploit opportunities and adjust as warranted. Share lessons learned.
- Be prepared to learn and adapt.

11.6 Conclusion

The study team applied the Profiler to an enterprise outside the information technology domain and, with minimum translation, was able to use it effectively to describe and synthesize the different facets of the environment in which this agency operates. Using the Profiler as an interview tool, the team discovered fundamental consistencies in how different groups of managers and stakeholders perceive the situation. These consistencies reflect the broad conditions that frame how this agency and the programs that it runs must operate. The team also found some unanticipated differences in perspective that relate to the respondents' roles in the organization. In retrospect, this makes intuitive sense, yet it was neither an anticipated nor even a sought-after outcome.

The resulting profile proved a useful way of prioritizing effort. In particular, the findings focused attention on those wedges that extended into the outermost ring. In this particular case, two wedges were consistently reported as being in the "messy frontier," and both dealt with the uncertain and diverse external environment.

The team was able to identify and develop a first-order description of three interrelated practices around the theme of addressing the uncertain and diverse external environment. Two of these practices—environmental scanning and stakeholder mapping and management—appear in the literature as components of strategic management. However, they are not typically practiced in the normal course of systems engineering or program/project management. The third practice—uncertainty management—is still emerging and lacks documented best practices.

Most important, senior leadership recognized the strategic value of these processes and accepted its role in both "owning" and executing them. At the end of the study, senior leadership initiated plans to establish an executive-level strategic council and implement these recommendations.

Endnotes

1. One of four Congressional staffers was also interviewed. His profile was reported separately and not included in this discussion. However, it is included in the composite profile.
2. See DeWeck, O., and C. Eckert, 2007. A Classification of Uncertainty for Early Product and System Design. *MIT Engineering Systems Division Working Paper Series.* February.
3. This material and the accompanying figure are adapted from work in process by Keith McCaughin of The MITRE Corporation (McCaughin and DeRosa, 2006).
4. Risk management is the overarching process that encompasses identification, analysis, mitigation planning, mitigation plan implementation, and tracking.

References

Ackoff, R. 1993. From Mechanistic to Social System Thinking. Paper presented at the *Systems Thinking in Action Conference*, November 1993. http://acasa.upenn.edu/socsysthnkg. pdf (accessed 10 February 2008).

Ackoff, R. 1994. *The Democratic Corporation: A Radical Prescription for Recreating Corporate America and Rediscovering Success*. New York: Oxford University Press.

Ackoff, R. 2004. Transforming the Systems Movement. Presented at the *Third International Conference on Systems Thinking in Management*, 26 March 2004. http://www.acasa. upenn.edu/RLAConfPaper.pdf (accessed 28 February 2008).

Albano, S. Undated. Lessons Learned in the Real World. Cambridge, MA: Auto-ID Center. http://islab.oregonstate.edu/koc/ece399/notes/rfid-field-test.pdf (accessed 7 December 2009).

Alberts, D.S., and R.E. Hayes. 2002. *Code of Best Practices for Experimentation*. Washington, DC: DoD Command and Control Research Program.

Amin, M. 2000. Modeling and Control of Electric Power Systems and Markets. *IEEE Control Systems Magazine*, August: 20–24. http://ieeexplore.ieee.org/iel5/37/18591/00856176. pdf (accessed 1 March 2008).

ANSI-EIA, American National Standards Institute/Electronic Industry Alliance. 1999. Processes for Engineering a System. ANSI/EIA-632-1998. ANSI/EIA.

Ashton. K. 2003. Testimony of Kevin Ashton, Executive Director, Auto-ID Center, before the California State Senate Subcommittee on New Technologies, Hearing on RFID and Privacy, 18 August 2003.

Auto-ID Center. 2002. Managing External Communications. Briefing presented 7 February 2002. http://cryptome.org/rfid/external_comm.pdf (accessed 24 February 2008).

Auto-ID Center. 2002. *Technology Guide*. Auto-ID Center, Cambridge, MA: Massachusetts Institute of Technology.

Auto-ID Labs. Undated. *History of the Auto-ID Center*. Cambridge, MA: Auto-ID Labs.

Bar-Yam, Y. 2003. When Systems Engineering Fails—Toward Complex Systems Engineering. *International Conference on Systems, Man & Cybernetics 2003*, 2: 2021–2028. Piscataway, NJ; IEEE Press. http://www.necsi.edu/projects/yaneer/E3-IEEE_final.pdf (accessed 1 March 2008).

Barabasi. A., et al. 2002. Scale-Free and Hierarchical Structures in Complex Networks, 25 November 2002. http://www.nd.edu/~networks/Publication%20Categories/03%20 Journal%20Articles/Physics/Scalefree-Hierarchical_Sitges%20Proceedings-Complex%20 Networks,%20.pdf (accessed 23 February 2007).

Boehm, B.W. 1988. A Spiral Model of Software Development and Enhancement. *Computer,* 5(21): 61–72.

Brock, D.L. 2001. Integrating the Electronic Product Code (EPC) and the Global Trade Number (GTIN). White Paper MIT-AUTOID-WH-004. Cambridge, MA: Auto-ID Center.

Brock, D.L. 2001. The Physical Markup Language: A Universal Language for Physical Objects. Cambridge, MA: Auto-ID Center.

Brock. D.L. 2003. Private conversation. Cambridge, MA: Auto-ID Center December 2003.

C/NET News.com. 2003. Wal-Mart Cancels 'Smart Shelf' Trial. 8 July 2003. http://www.news.com/2100-1017_3-1023934.html (accessed 24 February 2008).

Cady, A. 2003. Technologies for Enterprise Modernization, the MITRE Corporation, McLean, VA, briefing (30 September 2003).

Cantor, M. 2002. *Software Leadership: A Guide to Successful Software Development.* New York: Addison-Wesley.

Cebrowski, A.K. and J.J. Garstka. 1998. Network Centric Warfare: Its Origin and Future. *Proceedings of the Naval Institute,* 124(1): 28–35.

Charette, R.N. 2005. Why Software Fails. *IEEE Spectrum,* 42(9): 42–49. http://spectrum.ieee.org/sep05/1685 (accessed 23 February 2007).

Checkland, P.B. 1978. The Origins and Nature of "Hard" Systems Thinking. *Journal of Applied Systems Analysis,* 5: 99–100.

Checkland, P.B. 1981. *Systems Thinking, Systems Practice.* Chichester: Wiley.

Checkland, P.B. 1989. Soft Systems Methodology in Rosenhead, J. (Ed.), *Rational Analysis for a Problematic World,* Chichester: Wiley, pp. 71–100.

CIO Staff. 2004. GM's Cure for Complexity. *CIO Magazine,* 26 October 2004. www.cio.com.au/index.php/id;1706983620;fp;4;fpid;10 (accessed 23 February 2007).

Clark, S., K. Traub, D. Anarkat, and T. Osinski. 2003. Auto-ID Savant Specification 1.0. White Paper MIT-AUTOID-TM-003. Cambridge, MA: Auto-ID Center.

Collens, J.R., Jr., and R. Krause. 2005. *Theater Battle Management Core System Systems Engineering Case Study,* Air Force Institute of Technology, Center for Systems Engineering. https://acc.dau.mil/CommunityBrowser.aspx?id=37601 (accessed 1 March 2008).

Conklin, J. 2006. Wicked Problems and Social Complexity. CogNexus Institute. http://cognexus.org/wpf/wickedproblems.pdf (accessed 10 February 2008).

Cordesman, A.H. 2003. *The "Instant Lessons" of the Iraq War, Main Report.* Washington, DC: Center for Strategic and International Studies.

Courtney, H., J. Kirkland, and P. Viguerrie. 1995. Strategy under Uncertainty. *Harvard Business Review* November–December 1995. http://www.exchange.unisg.ch/org/lehre/exchange.nsf/1176ad62df2ddb13c12568f000482b94/5d58eda8f1f3dc90c125 6fa3005ebc52/$FILE/Courtney,%20Kirkland,%20Viguerie%20-%20Strategy%20 Under%20Uncertainty%20(McKinsey).pdf (accessed 29 February 2008).

CSTB, Computer Science and Telecommunications Board. 2000. *Making IT Better: Expanding Information Technology Research to Meet Society's Needs.* Washington, DC: National Academy Press.

Czerwinski, T. 1998. *Coping with the Bounds: Speculations on Nonlinearity in Military Affairs.* Washington, DC: National Defense University, Institute for National Strategic Studies.

Dahmann, J.S., and M. Crisp. 2003/2004. Joint Distributed Engineering Plant—Next Generation Infrastructure for Network Centric Engineering and Test. *The ITEA Journal of Test and Evaluation,* 24(4): 55–65.

DAU, Defense Acquisition University. 2005. *Glossary: A Glossary of Defense Acquisition Acronyms and Terms, 12th edition.* Ft. Belvoir, VA: Defense Acquisition University Press. http://www.dau.mil/pubs/glossary/12th_Glossary_2005.pdf (accessed 1 March 2008).

DAU, Defense Acquisition University. 2006. Defense Acquisition Guidebook. Version 1.06. Washington, DC: Defense Acquisition University.

De Meyer, A., C.H. Loch, and M.T. Pich. 2002. Managing Project Uncertainty. *Sloan Management Review,* 43(2):60–67.

De Neufville, R. 2004. Uncertainty Management for Engineering Systems Planning and Design, Engineering Systems Monograph. Presented at the *MIT Engineering Systems Symposium.* 29–31 March 2004. http://esd.mit.edu/symposium/pdfs/monograph/uncertainty.pdf (accessed 28 February 2008).

DeGrace, P., and L.H. Stahl. 1990. *Wicked Problems, Righteous Solutions.* Englewood Cliffs, NJ: Yourdon Press.

DHS, Department of Homeland Security. 2005. Fact Sheet: U.S, Department of Homeland Security FY 2006 Budget Request Includes Seven Percent Increase. Washington, DC. DHS Press Release.

DHS, Department of Homeland Security. 2005. National Plan to Achieve Maritime Domain Awareness for the National Strategy for Maritime Security. Washington, DC. www.dhs.gov/xlibrary/assets/HSPD_MDAPlan.pdf (accessed 23 February 2007).

DeWeck, O., and C. Eckert. 2007. A Classification of Uncertainty for Early Product and System Design. Cambridge, MA: MIT Engineering Systems Division Working Paper Series. http://esd.mit.edu/wps/2007/esd-wp-2007-10.pdf (accessed 27 February 2008).

Dillman, L.M. 2004. Radio Frequency Identification (RFID) Technology: What the Future Holds for Commerce, Security, and the Consumer. Witness testimony before the Subcommittee on Commerce, Trade, and Consumer Protection. 14 July 2004. http://worldcat.org/wcpa/top3mset/57299636 (accessed 24 February 2006).

DoD, Department of Defense. 2006. *Risk Management Guide for DoD Acquisition,* 6th Edition, Version 1.0, August, 2006.

DoD, Department of Defense. 2001a. Report on Network Centric Warfare, Sense of the Report. Submitted to the Congress. Washington, DC. www.dodccrp.org/files/ncw_report/report/ncw_sense.doc (accessed February 23, 2007).

DoD, Department of Defense. 2001b. Network Centric Warfare, Report to Congress. Washington, DC.

DoD, Department of Defense. 2003a. *Risk Management Guide for Department of Defense Acquisitions.* Belvoir, VA: Defense Acquisition University Press.

DoD, Department of Defense. 2003b. *Transformation Planning Guidance.* Washington, DC. www.oft.osd.mil/library/library_files/document_129_Transformation_Planning_Guidance_April_2003_1.pdf (accessed 23 February 2007).

Duce, H. 2003. Executive Briefing, Public Policy: Understanding Public Opinion. Cambridge, MA: Auto-ID Center. www.autoidlabs.org/single-view/dir/article/6/199/page.html (accessed 23 February 2007).

Dutchyshyn, H. 2005. Implementing the JBC2 Roadmap: A JSSEO Perspective. Presentation at the *NDIA Conference on Net Centric Operations, Interoperability, and Systems Integration.* http://www.dtic.mil/ndia/2005netcentric/2005netcentric.html (accessed 9 December 2009).

Dvir, D., A.J. Shenhar, and S. Alkaher. 2003. From a Single Discipline Product to a Multidisciplinary System: Adapting the Right Style to the Right Project. 2. *Systems Engineering,* 6(3): 123–134.

Dvir, D., S. Lipovetky, A. Shenhar, and A. Tishler. 1998. In Search of Project Classification: A Non-Universal Approach to Project Success Factors. *Research Policy,* 27: 915–935.

EAN UCC. November 1999. *White Paper on Radio Frequency Identification.* EAN International.

EPCglobal. 2004. EPC Tag Data Standards Version 1.1 Rev 1.24 Standards. EPC Global Standards Specification. http://www.epcglobalinc.org/standards/tds/tds_1_1_rev_1_27-standard-20050510.pdf (accessed 27 February 2008).

Faughn, A.W. 2002. Interoperability: Is It Achievable? Cambridge, MA: Harvard University Program on Information Resources Policy. P-02-6. http://www.pirp.harvard.edu/pubs_pdf/faughn/faughn-p02-6.pdf (accessed 1 March 2008).

Federowicz, J., J.L. Gogan, and C.B. Williams. 2006. The E-Government Collaboration Challenge: Lessons from Five Case Studies. Washington, DC: IBM Center for the Business of Government.

Floerkemeier, C., D. Anarkat, T. Osinski, and M. Harrison. 2003. PML Core Specification 1.0. Auto-ID Center Recommendation.

Fosberg, K., H. Mooz, and H. Cotterham. 2000. *Visualizing Project Management, 2nd edition.* New York: John Wiley & Sons.

Frank, M. 2000. Engineering Systems Thinking and Systems Thinking. *Systems Engineering,* 3(3): 163–168.

Frank, M. 2002. What Is "Engineering Systems Thinking?" *Kybernetes,* 31(9/10): 1350–1360.

Friedman, G., and A.P. Sage. 2004. Case Studies of Systems Engineering and Systems Acquisition. *Systems Engineering,* 7(1): 84–97.

Gaffin, A. 2007. Homeless Man Disrupts Internet 2 Service. 2 May 2007. *Network World.* http://www.networkworld.com/news/2007/050207-internet2-fire.html (accessed 27 February 2008).

Gansler, J.S. 2000. Memorandum for Secretaries of the Military Departments. Subject: Implementation Guidance for the Single Integrated Air Picture (SIAP) System Engineering (SE) Task Force (26 October, 2006).

Gilder, G. 1993. Metcalf's Law and Legacy. *Forbes ASAP.* www.seas.upenn.edu/~gaj1/ggindex.html (accessed 28 February 2008).

GS1 US. 2006. The Universal Product Code. www.uc-council.org/upc_background.html (accessed 28 February 2008).

Hall, A.D. 1962. *A Methodology for Systems Engineering.* Princeton, NJ: D. Van Nostrand.

Harvard Business Review staff. 1999. *Harvard Business Review on Managing Uncertainty.* Cambridge, MA: Harvard Business School Press.

Hastings, D., and H. McManus. 2004. A Framework for Understanding Uncertainty and its Mitigation and Exploitation in Complex Systems. Presented at *MIT Engineering Systems Symposium.* Cambridge, MA. http://ocw.mit.edu/NR/rdonlyres/Engineering-Systems-Division/ESD-85JFall-2005/695980D5-FDEE-4C4D-A505-14B8E9C95841/0/uncertainty.pdf (accessed 10 February 2008).

Highsmith, J.A. 2000. Adaptive Software Development: A Collaborative Approach to Managing Complex Systems. New York, NY: Dorset House.

Highsmith, J.A. 2004. *Agile Project Management: Creating Innovative Products.* Redwood City, CA: Addison-Wesley Longman Publishing.

Hillson, D. 2004. *Effective Opportunity Management for Projects.* Petersfield, Hampshire: United Kingdom Risk Doctor & Partners; New York: Marcel Dekker, Inc.

Hines, M. 30 April 2004. Wal-Mart Turns on Radio Tags. *ZDNet.* http://news.zdnet. com/2100-9584_22-5202240.html (accessed 24 February 2008).

Hughes. T.P. 1998. *Rescuing Prometheus: Four Monumental Projects that Changed the Modern World.* New York, NY: Pantheon.

IEEE, Institute of Electrical and Electronic Engineers.1990. IEEE Std 610.12 – 1990. *The IEEE Standard Glossary of Software Engineering Terminology.* New York, NY: IEEE.

IEEE, Institute of Electrical and Electronics Engineers. 1997. Std 100-1996, *The IEEE Standard Dictionary of Electrical and Electronics Terms, sixth edition.* New York, NY: IEEE.

INCOSE, International Council on Systems Engineering. 2004. What Is Systems Engineering? http://www.incose.org/practice/whatissystemseng.aspx. (accessed 5 December 2009).

Intelligence Reform and Terrorism Prevention Act of 2004, Public Law No. 108-458, 118 Stat 3638.

ISO/IEC, International Standards Organization/International Electrotechnical Commission. 2002. Systems Engineering—System Life Cycle Processes. Geneva, Switzerland: ISO/ IEC.

Jackson, M.C., and P. Keys. 1984. Towards a System-of-Systems Methodologies [sic], *Journal of the Operations Research Society,* 35(6): 473–486.

JTAMDO, Joint Theater Air and Missile Defense Organization). 1996. *2010 Battle Management Concept for Joint Theater Air and Missile Defense.*

JTAMDO, Joint Theater Air and Missile Defense Organization. 1998. *Capstone Requirements Document for Theater Missile Defense.*

Keating, C., et al. 2003. System of Systems Engineering. *Engineering Management Journal,* 15(3): 36.

Kirsner, S. 2002. Building a 'Radar for Everyday Products': B2B: A New Technology Reinvents the Bar Code to Track Goods and Change the Basics of Retail. *Newsweek,* 18 March 2002.

Krikeles, B., R. Merenyi, and J. Brtis. 17 May 2004. Use of MDA in the SIAP Program. Presentation given at the *MDA Implementers' Workshop,* 17–20 May 2004. Orlando, FL.

Krygiel, A.J. 1999. *Behind the Wizard's Curtain: An Integration Environment for a System-of-Systems.* Washington, DC: National Defense University, Institute for National Strategic Studies. CCRP Publication Series.

Landt, J. 2001. Shrouds of Time, The History of RFID. Pittsburgh, PA: Association for Automatic Identification and Data Capture Technologies. Available online at: www. aimglobal.org/technologies/rfid/resources/shrouds_of_time.pdf.

Larmon, C., and V.R. Basili. 2003. Iterative and Incremental Development: A Brief History. *Computer,* June: 47–56. http://www2.umassd.edu/SWPI/xp/articles/16047.pdf (accessed 26 February 2008).

Leiner, B.M., et al. 2003. A Brief History of the Internet. V3.32. Internet Society. Available online at: www.isoc.org/internet/history/brief.shtml.

Leong, K.S., M.L. Ng, and D.W. Engels. 2006. EPC Network Architecture. Auto-ID Network Labs Working Paper SWNET-12. Cambridge, MA: Auto-ID Center.

Loch, C.H., A. DeMeyer, and M.T. Pich. 2006. *Managing the Unknown: A New Approach to Managing High Uncertainty and Risk in Projects.* Hoboken, NJ: John Wiley & Sons.

Loch, C.H., M.E. Solt, and E. Bailey. 2005. Diagnosing and Managing Unforeseen Uncertainty to Improve Venture Capital Returns. INSEAD Working Paper 2004/1-1/ TOM, Fontainebleau: INSEAD. http://flora.insead.edu/insead/jsp/system/win_main. jsp (accessed 10 February 2008).

Lulay, D. 2003. The Future Impact of RFID Technologies on Supply Chain Management Strategies. Presentation to *RosettaNet Board Meeting*.

Magrassi, P., and T. Berg. 2002. A World of Smart Objects: The Role of Auto-Identification Technologies. Strategic Analysis Report, R-17-2243. 12 August 2002. Stamford, CT: Gartner Group.

Maier, M. 1996. Architecting Principles for Systems of Systems. Proceedings of the *Sixth Annual International Symposium, International Council on Systems Engineering*, pp. 567–574. Boston. www.infoed.com/Open/PAPERS/systems.htm. (accessed 1 March 2008).

Martin, J.N. 1996. *Systems Engineering Guidebook: A Process for Developing Systems and Products*. Boca Raton, FL: CRC Press.

Matthews, D., M. Burke, and P. Collier. 2000. Core Concepts of Joint Systems. *Proceedings of the SETE2000 Conference*. 15–17 November 2000. Brisbane, Australia. http://www.seecforum.unisa.edu.au/SETE2000/ProceedingsDocs/45P-Matthews.PDF (accessed 1 March 2008).

McCaughin, K.L., and J.K. DeRosa. 2006. Stakeholder Analysis to Shape the Enterprise. Presented at the *International Conference on Complex Systems (ICCS) 2006*, 25–30 June 2006, Boston, MA.

McGrath, R., and I.C. MacMillan. 1995. Discovery-Driven Planning. *Harvard Business Review*, July–August: 44–54.

Miller,W. 2004. Cursor on Target. *Military Information Technology Online*, 8(7). http://www.military-information-technology.com/article.cfm?DocID=596 (accessed 1 March 2008).

Mooz, H., and K. Forsberg. 1996. The Importance of Systems Engineering. Herndon, VA: Center for Systems Management, Inc.

NASCIO, National Association of State Chief Information Officers. 2007. Connecting State and Local Government: Collaboration through Trust and Leadership. Lexington, KY: NASCIO.

National Commission on Terrorist Attacks upon the United States (The 9/11 Commission). 2004. *The 9/11 Commission Report. Executive Summary*. Washington, DC: Government Printing Office. Available at: http://www.gpoaccess.gov/911/pdf/execsummary.pdf.

National Security Council. 2007. *National Strategy for Information Sharing*. Washington, DC. http://georgewbush-whitehouse.archives.gov/nsc/infosharing/index.html.

OSAF, Office of the Secretary of the Air Force. 2000. Air Force Policy Directive (AFPD) 63-12, "Assurance of Operational Safety, Suitability, & Effectiveness," and Air Force Instruction (AFI) 63-1201. Washington, DC. www.e-publishing.af.mil/shared/media/epubs/AFPD63-12.pdf (accessed 1 March 2008).

O'Shea, M., and A. Bigornia. 2004. Presentation at *EPC Global U.S. Conference*. 30 September 2004.

Owens, W.A. 1996. The Emerging US System-of-Systems. *National Defense University Strategic Forum*. 63. February 1996. http://stinet.dtic.mil/cgi-bin/GetTRDoc?AD=A DA394313&Location=U2&doc=GetTRDoc.pdf (accessed 1 March 2008).

Pich, M.T., C.H. Loch, and A. De Meyer. 2002. On Uncertainty, Ambiguity and Complexity in Project Management. *Management Science*, 48(8): 1008–1023.

PMI. 2001. *Project Management Institute Fact Book, 2nd edition*. Upper Darby, PA: Project Managment Institute.

Reed, D.P. 2001. The Law of the Pack. *Harvard Business Review*, February: 23–24 Release. Available online at www.dhs.gov/xnews/releases/press_release_0613.shtm.

RFID Journal. 2003. Wal-Mart Draws Line in the Sand. *RDID Journal*, 11 June 2003. http://www.rfidjournal.com/article/view/462/1/1 (accessed 24 February 2008).

RFID Journal. 2004. Target Issues RFID Mandate. *RFID Journal,* 20 February 2004. http:// www.rfidjournal.com/article/view/802/1/1 (accessed 24 February 2006).

Ritter, H.W.J., and M.M. Webber. 1973. Dilemmas in a General Theory of Planning. *Policy Sciences,* 4: 155–169.

Roberti, M. 2003. Analysis: RFID—Wal-Mart's Network Effect. 15 September 2003. *CIO Insight.* http://www.cioinsight.com/c/a/Trends/Analysis-RFID-WalMarts-Network-Effect/3/ (accessed 24 February 2008).

Robinson, S. *New York Times,* September 1999, http://www.nytimes.com/1999/09/30/ business/cut-in-fiber-cable-disrupts-internet-traffic-nationwide.html?pagewanted=1

Royce, W.W. 1970. Managing the Development of Large Software Systems. *Proceedings, IEEE WESCON.* 26(August): 1–9.

Sage, A., and C. Cuppan. 2001. On the Systems Engineering and Management of Systems and Systems and Federations of Systems. *Information, Knowledge, and Systems Management,* 325–345.

Sage, A.P. 1992. *Systems Engineering.* New York: Wiley Interscience.

Sage, A.. 2005. System of Systems: Architecture Based Systems Design and Integration. George Mason University, 10 October 2005. http://ieeesmc2005.unm.edu/smc_key-note_sage.pdf

Saltzer, J.H. 1999. Coping with Complexity. Presentation at the *ACM Seventeenth Symposium on Operating Systems Principles.* web.mit.edu/Saltzer/www/publications/ Saltzerthumbnails.pdf (accessed 29 February 2008).

Saltzer, J.H. 2004. Complexity Revisited. Notes for MIT Course 6.033 Computer Science Engineering. Lecture 26. web.mit.edu/6.033/2004/wwwdocs/handouts/L26.2004.pdf (accessed 29 February 2008).

Sarma, S.E., S.A. Weis, and D.W. Engels. 2002. RFID Systems, Security & Privacy Implications. MIT White Paper. MIT-AUTOID-WH-014. Cambridge, MA: Auto-ID Center. http://www.autoidlabs.org/single-view/dir/article/6/122/page.html (accessed 24 February 2008).

Schwartz, J. 2004. Get Ready for RFID. 17 February 2004. VARBusiness. http://www.crn. com/it-channel/18831449 (accessed 24 February 2008).

Senge, P., 1990. *The Fifth Discipline: The Art and Practice of the Learning Organization.* New York: Doubleday. pp. 71–72.

Shenhar, A.J. 1998. From Theory to Practice: Toward a Typology of Project Management Styles. *IEEE Transactions on Project Management,* 45(1):33–48.

Shenhar, A.J. 2001. One Size Does Not Fit All Projects. *Management Science,* 47(3):394 414.

Shenhar, A.J., and D. Dvir. 1996. Toward a Typology Theory of Project Management. *Research Policy,* 25:607–632.

Shenhar, A.J., and D. Dvir. 2007. *Reinventing Project Management: The Diamond Approach to Successful Growth and Innovation.* Boston, MA: The Harvard Business School Press.

SIAP SETF, 2000. Charter for Single Integrated Air Picture System Engineer Task Force, Oct. 26, 2000.

SIAP SETF, 2001. Single Integrated Air Picture (SIAP) Block 0 Decision Support Binder.

SIAP SETF, Undated. Integrated Architecture Development. Briefing.

Sommer, S.C., and C.H. Loch. 2004. Selectionism and Learning in Projects with Complexity and Unforeseeable Uncertainty. *Management Science,* 50(10): 1334–1347.

Songini, M., 2007. Wal-Mart Shifts RFID Plans, *Computerworld,* Feb 26, 2007.

Songini, M.L. 2007. Wal-Mart Shifts RFID Plans. *Computer World.* www.computerworld. com/action/article.do?command=viewArticleBasic&articleId=284115 (accessed 23 February 2008).

Sterman, J.D. 2000. *Business Dynamics: Systems Thinking and Modeling for a Complex World.* New York: McGraw-Hill. p. 21.

Stevens, R., 2008. Profiling Complex Systems, Proceedings of the IEEE International Systems Conference, Montreal, Canada, April 2008.

Summer, S.C., and C.H. Loch. 2004. Selectionism and Learning in Projects with Complexity and Unforeseeable Uncertainty. *Management Science,* 50(10): 1334–1347.

Swedberg, C. 2004. States Move on RFID Privacy Issue. *RFID Journal,* 30 April, 2004. http://www.rfidjournal.com/article/view/924 (accessed 24 February 2008).

Texas Instruments. 2003. The Buzz About Emerging RFID. RFID in the News, 27 March 2003. www.ti.com/rfid/docs/news/in_the_news/2003/3-31-03.shtml (accessed 29 February 2008).

The White House. 2007. *National Strategy for Sharing Information:* Successes and Challenges in Improving Terrorism-Related Information Sharing. Washington, DC. October 2007. www.whitehouse.gov/nsc/infosharing/NSIS_book.pdf (accessed 29 February 2008).

USAF, U.S. Air Force. 2005. Report on System of Systems Engineering for Air Force Capability Development. SAB-TR-05-04.

Valerdi, R., A.M. Ross, and D.H. Rhodes, 2007. A Framework for Evolving System of Systems. *Crosstalk.* October 2007. www.stsc.hill.af.mil/Crosstalk/2007/10/0710Valer diRossRhodes.html (accessed 29 February 2008).

Wal-Mart, 2009. Electronic Product Codes (EPC) – Helping Wal-Mart Help Customers and Suppliers. walmartstores.com/download/2449.pdf Accessed 5 March 2010.

Webster's New World Dictionary, Second College Edition, 1996. New York, NY: John Wiley & Sons.

White, B.E. 2006a. On the Pursuit of Enterprise Opportunities by Systems Engineering Organizations. Presentation to the *IEEE International Conference on System of Systems Engineering,* 24–26 April 2006, Los Angeles, CA.

White, B.E. 2006b. Enterprise Opportunity and Risk, Presentation to the *INCOSE 2006 Symposium,* 9–13 July 2006, Orlando, FL.

Woods, J., and A. White. 2003. September EPC Symposium Shows RFID Is Still Not Ready. 31 October 2003. Gartner. Research Note E-21-1250.

Index

Milton Keynes UK
Ingram Content Group UK Ltd.
UKHW031130141024
449569UK00006B/293